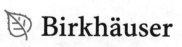

Hossein Rastgoftar

Continuum Deformation of Multi-Agent Systems

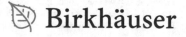 Birkhäuser

Hossein Rastgoftar
Department of Aerospace Engineering
University of Michigan Ann Arbor
Ann Arbor, MI, USA

ISBN 978-3-319-82391-1   ISBN 978-3-319-41594-9   (eBook)
DOI 10.1007/978-3-319-41594-9

Mathematics Subject Classification (2010): 93C10, 93C35, 93C95, 70E60

Printed on acid-free paper

This book is published under the trade name Birkhäuser, www.birkhauser-science.com
The registered company is Springer International Publishing AG
The registered company address is: Gewerbestrasse 11, 6330 Cham, Switzerland

To my Parents

Preface

Formation control in MASs has received a great deal of attention during the past two decades. Formation control has found different applications including formation flight, transportation engineering, air traffic control, missions in hazardous environment, environmental sampling, and gaming. Some advantages of collaboration among the agents and preserving a formation are reducing cost of the system, increasing robustness and efficiency of the system, and having better fault tolerance, structural flexibility, and capability of reconfiguration. The most recent algorithms for formation control are consensus algorithm, partial differential equation (PDE)-based, and containment control. These approaches, which have been inspired by heat diffusion problems, are interesting because they apply Laplacian control to achieve a global coordination among the agents through local interagent communication.

I started my PhD research by studying features of these three promising methods. After coming to a deep understanding of the fundamentals of these approaches, I attempted to find solutions for the following three problems:

Avoidance of interagent collision: The consensus model and the containment control method can theoretically address stability of evolution of an MAS when all communication weights are positive. However, avoidance of interagent collision among the agents is not necessarily assured, during a transition, for a defined fixed interaction topology with a positive set of communication weights.

Rigidity of MAS formation: For motion control applications when an MAS is applying the consensus model, interagent distances in the desired formation asymptotically converge to constant values. This will result in rigidity of the desired formation, and therefore, collective motion of agents may be difficult where passing through a narrow channel is required.

Containment during evolution: Followers of an MAS evolving under the PDE-based approach or containment control model may leave the containment region, prescribed by the leaders, during evolution (transition), although they are ultimately placed inside the convex hull which is defined by leaders. Thus, collision with obstacles may not be avoided.

It has been demonstrated how these problems can be addressed by treating motion control as continuum deformation. A leader-follower algorithm that is based on continuum mechanics principles can solve the aforementioned. Under a continuum deformation, no two particle agents occupy the same position during evolution, while the MAS has the capability of large expansion and compression. As a result, interagent collision can be avoided. A specific class of deformation mappings that are called homogeneous is considered, where the Jacobian of the mapping is only a function of time and is not spatially varying. A homogeneous transformation of an MAS in an n-D motion space can be uniquely related to trajectories of $n + 1$ leader agents, where leaders' trajectories are chosen such that collisions with obstacles are avoided. Then follower agents can acquire the desired homogeneous deformation map (prescribed by leaders) either 1) through no interagent communication by knowing leaders' positions in a finite horizon of time or 2) local communication with some adjacent agents and applying different communication protocols. Additionally, it can be shown how an arbitrary distribution of an MAS can be deployed in any desired formation through local communication (with communication weights that are determined based on positions of agents in the desired configuration) where avoidance of interagent collision during evolution is properly addressed.

Also, the issues of robustness to communication failure and asymptotic tracking of desired positions can be addressed when an MAS evolving in an n-D motion space applies either fixed or switching communication topologies. Furthermore, the effect of communication delays in an MAS evolving under consensus algorithms or homogeneous maps is another interesting topic. For an MAS containing a large number of agents, the order of the dynamics of MAS evolution is high, and therefore, available methods for stability analysis of delayed systems may be inefficient. To deal with this issue, a formulation for maximum allowable communication delay on the basis of Eigen-analysis of the network Laplacian matrix will be proposed.

I would like to express my deepest gratitude to my parents for their unfailing love, encouragement, and supports. I would also like to thank my sister and brothers. They were always supporting me with their best wishes.

I would never have been able to finish this work without the unsurpassed knowledge and generous and inspiring supports of my beloved advisor, the late Professor Suhada Jayasuriya. I am so proud of being his last student. I will never forget the life lessons that I learned from him. He was a real gentleman with a very kind soul. I am still very saddened by missing such a great father.

I would like to gratefully and sincerely thank my advisor Professor Harry G. Kwatny, for his excellent guidance and support. It is really my great life honor being his student. It was really a great opportunity for me to learn from his immense knowledge.

I take this opportunity to express my sincere and heartfelt gratitude to Professor Mojtaba Mahzoon for what I learned from him at Shiraz University. I believe what I learned has been much more than just science and math. He truly has an important role in shaping my academic life.

Ann Arbor, MI, USA   Hossein Rastgoftar
January 2016

Contents

Chapter 1
Introduction

Formation control has received considerable attentions during the past two decades. Some applications like formation flight, transportation engineering, air traffic control, gaming, maneuvering in a hazardous environment, and environmental sampling have been listed in literature for formation control. Formation control in a multi-agent system (MAS) has many advantages [91]. For example, keeping formation increases robustness and efficiency of a system reduces the cost of a system, and results in better fault tolerance and capability of reconfiguration [6, 8, 9, 19, 140].

1.1 Available Methods for Formation Control

Common approaches for formation control in MASs are the leader follower [8, 13, 18, 21, 22, 24–29, 32, 38, 44, 49, 59, 60, 67, 81, 128, 129, 132, 141, 143, 163, 164, 173], the virtual structures [52, 62, 66, 79, 82, 118, 138, 144, 157, 165–167], the artificial potential functions [1, 2, 30, 40, 41, 43, 55, 84–86, 100, 103, 120, 123–126, 148, 154, 159, 161, 162, 168, 172], the behavioral based [4, 5, 7, 23, 76, 121], the consensus algorithm [10, 16, 20, 53, 57, 69, 71, 72, 74, 77, 78, 92, 93, 98, 102, 117, 119, 127, 134, 136, 139, 147, 150, 151, 155, 156, 158, 160], the partial differential equation (PDE) based method [36, 37, 39, 42, 61, 80, 83], and the containment control [14, 15, 34, 54, 73, 145]. Features of these available methods have been recently investigated in the author's MS dissertation [104].

In this chapter, the recent methods for the collective motion of a multi-agent system including the consensus algorithm, the PDE-based method, and the containment control are investigated. These three methods are inspired by heat diffusion problems [34]; they apply Laplacian control to achieve a global coordination among the agents through local interagent communication.

© Springer International Publishing AG 2016
H. Rastgoftar, *Continuum Deformation of Multi-Agent Systems*,
DOI 10.1007/978-3-319-41594-9_1

1.1.1   The Consensus Algorithm

Applying the consensus algorithm requires an MAS to use an irreducible or a completely triangularly reducible graph for interagent communication [102]. Then, it is shown that network transient state cooperatively comes to a consensus state, if the communication weights are all positive. Mathematically speaking, consensus state is obtained if entries of the transient state vector all converge to a unique desired value. Consensus state depends on (i) the initial value of the network state, and (ii) left and right eigenvectors of the Laplacian matrix of the communication graph.

So far, different applications such as motion control [117, 119, 127, 134], network clock synchronization [16], distributed sensing [95], medical application [45, 92, 116], and power systems [10, 77, 152, 158, 160, 171, 174] have been suggested for the consensus model. In this regard, a couple of interesting issues corresponding to the distributed convergence under consensus models have been addressed by the researchers. In references [69, 93, 150, 151, 156], it has been shown how an MAS applying switching communication topologies can asymptotically reach the consensus agreement. Furthermore, robustness of distributed convergence in MASs under communication failure [3, 57, 98, 170], sensor failure [56, 96, 133], actuator failure [51, 149, 178], and model uncertainty [11, 53, 72, 101, 169] are important issues that have been addressed by researchers. Coming to a consensus agreement under stochastically switching topologies was also developed in references [57, 74, 139]. The necessary and sufficient conditions for convergence of an MAS under consensus model, when the communication graph is generated by an ergodic and stationary random process, were developed in reference [136]. Stability analysis of MAS evolution under consensus algorithm with time delay has been another interesting field of research [20, 71, 78, 147, 155]. Evolution of multi-agent systems under discrete-time consensus algorithms is studied in [70, 75, 137]. Some interesting analytical models for stability analysis of retarded consensus problems are proposed in [12, 65, 89, 90]. In [12], the stability of a first order average consensus problem in presence of either constant or time-varying communication delays are analyzed, and upper bounds for the communication delays of each agent are determined based on the topology of the interagent communication. In [89, 90], the evolution of an MAS under first order consensus algorithm with and without self-delay is investigated, and an accurate set-valued Nyquist condition is provided for the communication delay of each agent through Eigen-analysis of the graph Laplacian matrix. Stability of the first order consensus problem with an arbitrary symmetric and bidirectional communication topology is achieved by introducing interconnection symmetries and using convexification in the complex plane [65].

1.1.2   The PDE-Based Model

The PDE-based method is another appealing approach which has been suggested for collective motion of an MAS. Under this setup, evolution of an MAS is usually modeled by a first order or second order PDE with spatially varying parameters. Agents of the MAS are then categorized as (i) leaders and (ii) followers, where leader agents are the agents placed at the boundary. Leaders' positions are prescribed by the imposed boundary conditions. The interior agents are considered as the followers where each follower communicates with some neighboring agents, and the communication weights are determined by discretization of the spatially varying part of the PDE. In [46, 47, 63, 97, 142], it is shown how an MAS using PDE-based methods can maintain a 1-D rigid formation. In reference [36, 37, 42], it was shown how a random distribution of agents can be stably deployed on desired planar curves, where the collective motion of the agents is governed by a reaction-advection-diffusion class of PDEs. In reference [83], it was shown how a collective motion of an MAS, prescribed by a nonlinear PDE, can be deployed in a desired formation through local communication. In reference [61], a PDE-based model reference adaptive control algorithm has been proposed for collective motion of an MAS facing uncertain heterogeneous interagent communication. Moreover, the PDE-based technique has been recently applied in wind-integrated power systems to control the inter-area oscillations of the generations connected to the transmission line [39, 80]. Furthermore, controlling load distribution in smart grids using PDE-based method is studied in [68, 87].

1.1.3   The Containment Control Approach

The containment control is a leader-follower model, where (i) leaders move independently and guide evolution of the agents toward a desired target; and (ii) followers apply consensus algorithm to reach the global coordination inside the containment region that is defined by the leaders [14, 15, 34, 54, 73]. In references [34, 54], hybrid control strategies (Go-Stop) are applied by the leaders to guide followers to a desired containment region. Containment control of a formation of an MAS where followers apply the double integrator kinematic model was presented in [15]. Containment control of an MAS applying either stationary or switching communication topologies was proposed in [14, 50, 135]. Moreover, containment control with followers applying complex communication weights was presented in [145]. Containment control of multi-agent under directed communication graphs is demonstrated in [31, 50, 176]. Finite-Time containment control of multi-agents systems with second order dynamics is studied in [48]. Discrete-time containment control of multi-agent systems is demonstrated in [88, 175]. In [122, 146, 153], containment control of multi agent systems under higher order dynamics is investigated.

1.1.4 Problems Definitions

Although the consensus algorithm, the PDE-based method, and the containment control have very interesting features and are widely applied in different fields of engineering and science, there are some issues that have not been clearly addressed yet. For example, when the consensus algorithm with fixed communication weights is used for controlling of motion of an MAS, interagent distances in the desired formation asymptotically converge to some constant values. This will result in the rigidity of the desired formation, and therefore, collective motion of agents may be difficult in cases where passing through a narrow channel is required. In addition, requirements of the consensus algorithm and containment control method, for asymptotic distributed convergence, do not necessarily avoid interagent collision among the agents. To reach a desired formation by applying the consensus algorithm and containment control, the irreducibility of the interagent communication graph and positiveness of communication weights are the requirements [102]. However, for motion control applications, consistency of communication weights with the initial placement of agents is very important, that if ignored, agents may collide during evolution. For example, suppose an MAS applying the containment control method to evolve in a plane, where the initial positions of the agents and interagent communication graph are shown in Fig. 1.1. As it is seen, agents 1, 2, and 3 are the leaders; and agents 4, 5, and 6 are the followers. Leader agents all move independently toward the right with the same velocities $\frac{1}{3}m/s$. Follower agents apply the first order dynamic model presented in [14, 15, 34, 54, 73] to update their current positions based on the positions of the adjacent agents. Positive communication weights $w_{4,2} = w_{4,5} = w_{4,6} = \frac{1}{3}$, $w_{5,1} = w_{5,4} = w_{5,6} = \frac{1}{3}$, $w_{6},3 = \frac{1}{2}$ and $w_{6,4} = w_{6,6} = \frac{1}{4}$ are selected by followers. Thus, requirements for the containment control method are satisfied, and it is expected that followers ultimately converge to the containment region, when leaders settle. This is because (i) communication weights applied by the followers are all positive, and (ii) the interagent communication graph shown in Fig. 1.1 is irreducible. However, as illustrated in Fig. 1.1, agents 4 and 5 collide at the time $t = 0.45s$ because communication weights are not consistent with the agents' initial positions. Consequently, MAS evolution practically stops at the time t = 0.45, although, the asymptotic convergence of the transient MAS formation to a desired configuration could be theoretically achieved.

   In fact, the containment control and the PDE-based are conceptually very similar. They are leader-follower models for the distributed control in an MAS that guarantee achieving a desired global coordination through local interagent communication. Under these setups, leaders guide the motion of the agents toward the desired target, and followers update their positions through local communication. It can be assured that followers finally converge to the desired positions inside the containment region defined by the leaders' positions.

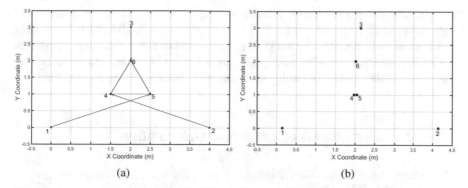

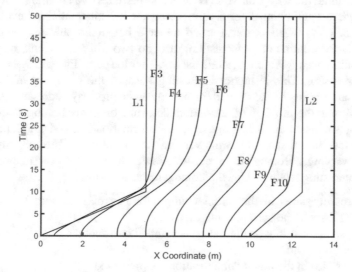

Fig. 1.1 (a) Initial distribution and interagent communication graph evolve in a plane; (b) interagent collision at the time $t = 0.45s$

Fig. 1.2 Schematic of leaving containment region when follower agent apply PDE-based model to evolve

Although the problem of asymptotic convergence of the transient positions of the followers to some desired positions inside the final containment control is addressed by applying the PDE-based and the containment control methods, it is not necessarily assured that followers remain inside the containment region throughout MAS evolution. In other words, there is no suggestion for determining the lower limits for the control gains applying by the followers in order to assure that followers remain inside the containment region during MAS evolution. In Fig. 1.2 evolution of an MAS applying the PDE-based method to evolve in a straight line is shown. In Fig. 1.2, the horizontal axis denotes position (m) and the vertical axis denotes time (s). As seen in Fig. 1.2, the leader agents 1 and 2 guide collective motion of the MAS, where they eventually settle at $x_{1,F} = 5m$ and $x_{2,F} = 12m$ at the time

$t = 10s$. One can easily observe that follower agents 3 and 5 leave the containment region during the time period $t \in [7,12]s$, although they finally reach the desired positions that are governed by the PDE (containment region is the interior points of the leading line segment whose end points are occupied by the leaders 1 and 2). This may result in the collision of follower agents with obstacles in the motion field.

1.2   MAS Evolution As Continuum Deformation

To address the aforementioned issues, a continuum based framework for the evolution of MAS in $\mathbb{R}^n$ have been recently proposed in [104, 106–113]. Note that for motion control applications, n denotes dimension of the motion space that can be 1, 2, or 3. Under this setup, an MAS is treated as particles of a continuum deforming in $\mathbb{R}^n$, where continuum deformation prescribes the kinematics of evolution of the agents. If an MAS is transformed under a homeomorphic mapping that is admissible for continuum deformation, then no two different agents occupy the same position during evolution, while the MAS has the capability of large expansion and compression. This is highly interesting because the problems of interagent collision and rigidity of the desired formation can be properly addressed.

In this book the problem of continuum deformation under homogeneous transformation is considered. A homogeneous transformation is a nonsingular mapping with the Jacobian that is only time-varying (the Jacobian of the mapping is not spatially varying). Homogeneous deformation is in fact a linear mapping with several interesting features. Under a homogeneous deformation,

- two crossing lines in the initial configuration is mapped into two different crossing lines in the current configuration,
- an ellipse in the initial configuration is mapped into another ellipse in the current configuration, and
- volume ratios in the initial configuration are preserved.

A leader-follower model for the evolution of an MAS under the homogeneous transformation has been recently proposed by the author. It is shown how a desired homogeneous mapping in $\mathbb{R}^n$ can be uniquely related to the desired trajectories chosen by $n+1$ leader agents, where leaders occupy vertices of a polytope in $\mathbb{R}^n$, called *leading polytope*. Notice that for MAS evolution problem in a 1-D space two leaders are placed at the end points of a leading line segment, for MAS evolution in a 2-D space, three leaders occupy vertices of a leading triangle, and for MAS evolution in a 3-D, four agents are considered as the leaders and placed at the vertices of a leading tetrahedron. Therefore, a desired homogeneous deformation can be designed by choosing proper trajectories for the leaders, where collisions of agents with obstacles in the motion field are avoided. Followers are the agents that are distributed inside the convex hull of the leading polytope. They can learn a desired homogeneous mapping, prescribed by the leaders, through (i) no interagent communication [104, 111, 113], (ii) local communication [104, 106–113], and

(iii) local perception. Additionally, principles of continuum mechanics are used to deploy an arbitrary distribution of agents on a desired formation in $\mathbb{R}^n$ [109, 112].

In Chapter 2, the idea of MAS evolution under no communication is presented. It is shown how followers can acquire a desired homogeneous mapping only by knowing leaders' positions. The proposed approach is scalable meaning that there is no restriction on the total number of followers acquiring homogeneous deformation. In addition, the MAS has the capability of enlargement and contraction, while interagent collision and collision with obstacles are properly addressed. Several interesting problems are considered in Chapter 2. These problems include

- homogeneous deformation of agents with and without constrained dynamics,
- homogeneous transformation of agents with linear dynamics,
- specifying an upper bound for the acting forces required for the, and
- obtaining optimal trajectories for the leaders minimizing the acceleration norms MAS evolution map defined by a homogeneous transformation, where agents are treated as particles of the Newtonian fluid.

Notice that considering MAS as particles of the Newtonian fluid imposes two holonomic constraints (resulting from the continuity and Navier-Stokes constitutive equations) on the leaders' motion.

In Chapter 3, a paradigm is presented for followers to acquire a desired homogeneous deformation through local communication. Two different communication protocols will be developed in this chapter: (i) minimum interagent communication, and (ii) polyhedral communication protocol.

Under minimum interagent communication protocol, each follower updates its position through interaction with $n + 1$ local agents, where communication weights are chosen such that they are consistent with initial positions of the agents. It will be shown how communication weights are uniquely specified based on initial positions of the agents, if every follower interacts only with $n + 1$ local agents. This is why this protocol is called minimum interagent communication. Applying minimum interagent protocol assures that followers asymptotically reach the desired states defined by a homogeneous deformation in $\mathbb{R}^n$.

A polyhedral communication protocol will also be proposed for followers to learn a desired homogeneous mapping prescribed by $n + 1$ leaders. Under this setup, each follower i is allowed to communicate with $p_i \geq n + 1$ local agents. Consequently, rate of convergence of MAS evolution can be improved because followers are permitted to increase their communication. Here, each follower continuously attempts to acquire the desired homogeneous mapping through preserving some volumetric weight ratios specified based on the agents' positions.

In Chapter 3, several further interesting problems, corresponding to the minimum interagent communication protocol, are investigated. These problems are outlined as follows:

Determining an upper limit for the followers' deviations from the desired positions defined by a homogeneous transformation: Since follower agents acquire desired positions prescribed by homogeneous deformation through local communication, they deviate from these desired positions during MAS evolution. However, an upper bound can be determined for the followers' deviations, and thus avoidance of interagent collision can be fully addressed. This upper limit depends on (i) initial distribution of the agents, (ii) control gain applied by the followers, (iii) total number of agents which is denoted by N, (iv) maximum of the magnitudes of the velocities of the leaders, and (iv) the dimension of MAS evolution which is denoted by n.

New formulation for followers' allowable communication delays using Eigen-analysis: Stability analysis of MAS evolution in the presence of communication delay will be also investigated in Chapter 3. Cluster Treatment of Characteristic Roots (CTCR) [33, 94, 131] and Lyapunov-Krasovskii [35, 58, 99, 177] are the two available approaches for stability analysis of the delayed dynamic systems. However, applying these common approaches for analyzing stability of an MAS consisting of a large number of agents is difficult due to higher order dynamics of evolution of the agents. To resolve this difficulty, a new formulation for the allowable communication delay of each agent is derived. The upper bound for the communication delay depends on the control gain applied by follower agents, and one of the eigenvalues of the communication matrix. Notice that the proposed formulation can be efficiently applied to analyze stability of delayed consensus problems. It is noted that for evolution under the consensus algorithm, communication matrix is the same as the Laplacian matrix.

In Chapter 4 continuum deformation of an MAS under higher-order dynamics is presented. It is demonstrated how the transient error which is the difference between the actual and desired positions of each follower vanishes during transition. This is advantageous because it can be assured that followers do not leave the transient convex hull that is defined by the leaders' positions at any time t. For this purpose, each leader is permitted to choose finite polynomial of the order $((p-1) \in \mathbb{N})$ for its trajectory connecting two consecutive way points. Then, position of each follower is updated by a dynamics of the order $p \in \mathbb{N}$, and it is demonstrated how followers' deviations from the desired positions, defined by a homogeneous mapping, vanish during transition. This is interesting because followers only access to the state information of $n+1$ local agents, and they do not directly interact with leaders that define the desired homogeneous deformation. Furthermore, stability analysis in presence of heterogenous communication delays is described in Chapter 4. In this regard, upper bounds for communication delays are obtained by using Eigen-analysis.

In Chapter 5, it is shown how homogeneous deformation of an MAS can be achieved by keeping alignment among the agents. This is advantageous because preserving alignment only requires the direction information and not the exact locations. Hence, agents that were initially aligned can acquire a desired homogeneous deformation through preserving alignment, and they can evolve collectively based on their perceptions from their in-neighbor agents. The idea of MAS evolution under local perception is inspired by the fact that natural biological swarms do not perform peer-to-peer communication to follow a group behavior. Biological swarms evolve based on individual agents' perception. However, most, if not all, of engineering based techniques rely on peer-to-peer communication that requires each individual agent to know the exact state information of its neighbors. Therefore sensors with high accuracy are required to be applied by individual agents in order to measure exact state information of the neighboring agents.

In Chapter 6, a leader-follower model for deployment of an arbitrary random distribution of agents on any desired formation in $\mathbb{R}^n$ is developed, where avoidance of interagent collision is assured during MAS evolution. For this purpose, motion in $\mathbb{R}^n$ is decomposed into n separate 1-D motion problems that are guided by two $2n$ different leaders. Then, followers update the qth components of their positions through communication with two adjacent agents, where communication weights are fixed, uniquely determined based on positions of the agents in the final configuration.

In Chapter 7, collective motion of multi-agent systems over nonlinear surfaces is treated as continuum deformation. It is shown how an MAS can move collectively on an arbitrary curve in a 3-D motion space, where the motion is guided by two leaders and each follower interacts with two in-neighbor agents. Furthermore, continuum deformation of an MAS on 2-D surfaces is demonstrated, where three leaders prescribe the mapping and each follower interacts with 3 local agents to acquire the mapping through local communication.

Chapter 2
Homogeneous Deformation without Interagent Communication

In this chapter, basics of MAS evolution as continuum deformation are presented, where an MAS is treated as particles of a continuum deforming under a homogeneous transformation. It is shown how a homogeneous deformation is acquired by the agents via no interagent communication. In this regard, agents' desired positions, defined by a homogeneous deformation in $\mathbb{R}^n$, are uniquely specified by the trajectories chosen by $n+1$ leaders. Followers acquire the desired homogeneous mapping only by knowing leaders' positions. Evolution of the followers with nonlinear constrained dynamics under a homogeneous transformation is investigated in this chapter. Homogeneous transformation of an MAS containing agents with linear dynamics is also studied.

2.1 Homogeneous Transformation

A continuum (deformable body) is a continuous region in $\mathbb{R}^n$ (n=1,2,3) containing infinite number of particles with infinitesimal size [64]. A continuum deformation is defined by the mapping $r(R,t)$, where $r \in \mathbb{R}^n$ denotes current position of a material particle which was initially placed at $R \in \mathbb{R}^n$. Notice that positions of the agents at the initial time t_0 is called *material coordinate* and denoted by R. A schematic of a continuum deformation is shown in Fig. 2.1.

It is noted that the Jacobian of the continuum deformation that is denoted by

$$Q(R,t) = \frac{\partial r}{\partial R} \in \mathbb{R}^{n \times n}. \tag{2.1}$$

is nonsingular, where $Q(R,t_0) = I_n$ ($I_n \in \mathbb{R}^{n \times n}$) is the identity matrix and t_0 denotes the initial time. If the Jacobian matrix is only a function of time, then the continuum

© Springer International Publishing AG 2016
H. Rastgoftar, *Continuum Deformation of Multi-Agent Systems*,
DOI 10.1007/978-3-319-41594-9_2

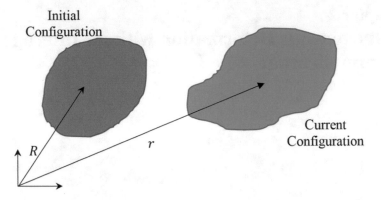

Fig. 2.1 Schematic of a continuum deformation

deformation is called *homogeneous transformation*. Therefore, a homogeneous transformation is defined by

$$r(t) = Q(t)R + D(t), \tag{2.2}$$

where $D(t) \in \mathbb{R}^n$ is called *rigid body displacement vector*.

2.1.1 Homogeneous Deformation of the Leading Polytope

Because homogeneous transformation is a linear mapping, elements of $Q(t)$ and $D(t)$ can be uniquely related to the components of the positions of $n + 1$ leaders. Suppose each leader agent is identified by a number $k \in V_L$ and $V_L = \{1, 2, \ldots, n + 1\}$. Leader agents are placed at the vertices of a leading polytope in $\mathbb{R}^n$, called *leading polytope*, therefore leaders' positions satisfy the following rank condition:

$$\forall t \geq t_0, Rank\left[r_2 - r_1 \ \ldots \ r_{n+1} - r_1\right] = n. \tag{2.3}$$

Let R and r in Eq. (2.2) be substituted by the leaders' initial and current positions, respectively, then the following $n + 1$ equations are obtained:

$$\begin{cases} r_1(t) = Q(t)R_1 + D(t) \\ r_2(t) = Q(t)R_2 + D(t) \\ \vdots \\ r_{n+1}(t) = Q(t)R_{n+1} + D(t) \end{cases} \tag{2.4}$$

r_k and R_k ($k \in V_L$) can be expressed with respect to the Cartesian coordinate system with the unit basis ($\hat{\mathbf{e}}_1$, $\hat{\mathbf{e}}_2$,..., $\hat{\mathbf{e}}_n$) as follows:

$$r_k = \sum_{q=1}^{n} x_{q,k}\hat{\mathbf{e}}_q \tag{2.5}$$

$$R_k = \sum_{q=1}^{n} X_{q,k}\hat{\mathbf{e}}_q, \tag{2.6}$$

where $X_{q,k}$ and $x_{q,k}$ are the qth components of the initial and current positions of the agent k. Then

$$\begin{cases} x_{q,1}(t) = \begin{bmatrix} X_{1,1} \ X_{2,1} \ \dots \ X_{n,1} \end{bmatrix} \begin{bmatrix} Q_{q1} \ Q_{q2} \ \dots \ Q_{qn} \end{bmatrix}^T + D_q \\ x_{q,2}(t) = \begin{bmatrix} X_{1,2} \ X_{2,2} \ \dots \ X_{n,2} \end{bmatrix} \begin{bmatrix} Q_{q1} \ Q_{q2} \ \dots \ Q_{qn} \end{bmatrix}^T + D_q \\ \vdots \\ x_{q,n+1}(t) = \begin{bmatrix} X_{1,n+1} \ X_{2,n+1} \ \dots \ X_{n,n+1} \end{bmatrix} \begin{bmatrix} Q_{q1} \ Q_{q2} \ \dots \ Q_{qn} \end{bmatrix}^T + D_q \end{cases} \tag{2.7}$$

It is noted that $q \in \{1, 2, \dots, n\}$, $D_q \in R$ is the qth entry of the vector $D \in \mathbb{R}^n$ and Q_{qj} is the qj entry of the Jacobian matrix $Q \in \mathbb{R}^{n \times n}$. The above set of $n+1$ equations can be simplified as

$$U_q = L_0 Q_q + D_q \mathbf{1}$$

where

$$U_q = \begin{bmatrix} x_{q,1} \ \dots \ x_{q,n+1} \end{bmatrix}^T \in \mathbb{R}^{n+1},$$

$$Q_q = \begin{bmatrix} Q_{q1} \ Q_{q2} \ \dots \ Q_{qn} \end{bmatrix}^T \in \mathbb{R}^n,$$

$$\mathbf{1} = \begin{bmatrix} 1 \ 1 \ \dots \ 1 \end{bmatrix}^T \in \mathbb{R}^{n+1},$$

$$L_0 = \begin{bmatrix} X_{1,1} & \dots & X_{n,1} \\ \vdots & & \vdots \\ X_{1,n+1} & \dots & X_{n,n+1} \end{bmatrix} \in \mathbb{R}^{(n+1) \times n}.$$

Therefore, entries of Q and D can be related to the components of the leaders' positions by

$$P_t = \begin{bmatrix} I_n \otimes L_0 \ I_n \otimes \mathbf{1} \end{bmatrix} J_t \tag{2.8}$$

where

$$J_t = \begin{bmatrix} Q_1^T & \dots & Q_n^T & D^T \end{bmatrix}^T \in \mathbb{R}^{(n+1)n} \tag{2.9}$$

$$P_t = \begin{bmatrix} U_1^T & \dots & U_n^T \end{bmatrix}^T \in \mathbb{R}^{(n+1)n}. \tag{2.10}$$

Followers' Distribution: Let each follower be identified by the index $i \in V_F$, where $V_F = \{n+2, \dots, N\}$. It is assumed that followers are all placed inside the leading polytope at the initial time t_0. Because leaders' positions satisfy the rank condition (2.3), $r_i(t)$ (position of the follower $i \in V_F$ at the time t) can be uniquely expanded as a linear combination of the leaders' positions as follows:

$$r_i(t) = r_1(t) + \sum_{k=2}^{n+1} p_{i,k}(t)(r_k(t) - r_1(t)) = (1 - \sum_{k=2}^{n+1} p_{i,k}(t))r_1(t) + \sum_{k=2}^{n+1} p_{i,k}(t)r_k(t)$$

$$= \sum_{k=1}^{n+1} p_{i,k}(t)r_k(t). \tag{2.11}$$

Notice that

$$\sum_{k=1}^{n+1} p_{i,k}(t) = 1. \tag{2.12}$$

By considering Eqs. (2.11) and (2.12), parameter $P_{i,k}(t)$ is uniquely obtained by solving the following set of linear algebraic equations:

$$\begin{bmatrix} x_{1,1} & x_{1,2} & \dots & x_{1,n+1} \\ x_{2,1} & x_{2,2} & \dots & x_{2,n+1} \\ \vdots & \vdots & \ddots & \vdots \\ x_{n,1} & x_{n,2} & \dots & x_{n,n+1} \\ 1 & 1 & \dots & 1 \end{bmatrix} \begin{bmatrix} p_{i,1} \\ p_{i,2} \\ \vdots \\ p_{i,n} \\ p_{i,n+1} \end{bmatrix} = \begin{bmatrix} x_{1,i} \\ x_{2,i} \\ \vdots \\ x_{n,i} \\ 1 \end{bmatrix}, \tag{2.13}$$

where x_{qj} denotes the qth component of the position of the agent $j \in \{i, 1, 2, \dots, n+1\}$. If leaders and followers deform under a homogeneous mapping, then parameters $p_{i,k}(i \in V_F$ and $k \in V_L)$ remain constant at any time $t \geq t_0$. Therefore, desired position of the followers $i \in V_F$, defined by a homogeneous deformation, is given by

$$r_{i,HT}(t) = \sum_{k=1}^{n+1} \alpha_{i,k} r_k, \tag{2.14}$$

where $\alpha_{i,k}$ is uniquely determined based on the initial positions of the follower i and $n+1$ leaders by solving the following set of linear algebraic equations:

$$\begin{bmatrix} X_{1,1} & X_{1,2} & \dots & X_{1,n+1} \\ X_{2,1} & X_{2,2} & \dots & X_{2,n+1} \\ \vdots & \vdots & \ddots & \vdots \\ X_{n,1} & X_{n,2} & \dots & X_{n,n+1} \\ 1 & 1 & \dots & 1 \end{bmatrix} \begin{bmatrix} \alpha_{i,1} \\ \alpha_{i,2} \\ \vdots \\ \alpha_{i,n} \\ \alpha_{i,n+1} \end{bmatrix} = \begin{bmatrix} X_{1,i} \\ X_{2,i} \\ \vdots \\ X_{n,i} \\ 1 \end{bmatrix}. \tag{2.15}$$

2.1.2   Homogeneous Deformation of the Leading Triangle

In this subsection homogeneous deformation of an MAS in a plane is considered. Let leader agents be placed at vertices of a triangle, called *leading triangle*, and followers be initially placed inside the leading triangle. Schematic of homogeneous deformation of an MAS in a plane is shown in Fig. 2.2. Because three leaders remain nonaligned, the rank condition (2.3) is simplified to

$$\forall t \geq t_0, Rank \left[r_2 - r_1 \ r_3 - r_1 \right] = 2. \tag{2.16}$$

Additionally, the desired position of the follower i, defined by a homogeneous deformation, is given by

$$r_{i,HT}(t) = \alpha_{i,1} r_1 + \alpha_{i,2} r_2 + \alpha_{i,3} r_3, \tag{2.17}$$

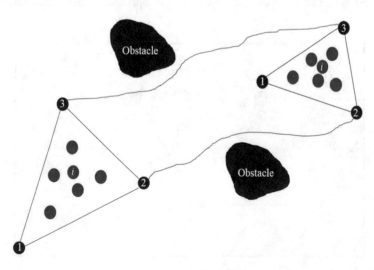

Fig. 2.2   Schematic of homogeneous deformation of an MAS in a plane

where $\alpha_{i,1}$, $\alpha_{i,2}$, and $\alpha_{i,3}$ are the unique solution of

$$\begin{bmatrix} X_1 & X_1 & X_3 \\ Y_1 & Y_2 & Y_3 \\ 1 & 1 & 1 \end{bmatrix} \begin{bmatrix} \alpha_{i,1} \\ \alpha_{i,2} \\ \alpha_{i,3} \end{bmatrix} = \begin{bmatrix} X_i \\ Y_i \\ 1 \end{bmatrix}. \tag{2.18}$$

Interpretation of parameters $\alpha_{i,k}$ ($i \in V_L$, $k \in V_F$): By solving set of linear algebraic equations in (2.18), parameters $\alpha_{i,1}, \alpha_{i,2}$, and $\alpha_{i,3}$ are obtained as follows:

$$\begin{cases} \alpha_{i,1} = \dfrac{X_i(Y_2 - Y_3) + Y_i(X_3 - X_2) + X_2 Y_3 - X_3 Y_2}{X_1(Y_3 - Y_2) + X_2(Y_1 - Y_3) + X_3(Y_2 - Y_1)} \\[2ex] \alpha_{i,2} = \dfrac{X_i(Y_3 - Y_1) + Y_i(X_1 - X_3) + X_3 Y_1 - X_1 Y_3}{X_1(Y_3 - Y_2) + X_2(Y_1 - Y_3) + X_3(Y_2 - Y_1)} \\[2ex] \alpha_{i,3} = \dfrac{X_i(Y_1 - Y_2) + Y_i(X_2 - X_1) + X_1 Y_2 - X_2 Y_1}{X_1(Y_3 - Y_2) + X_2(Y_1 - Y_3) + X_3(Y_2 - Y_1)} \end{cases} . \tag{2.19}$$

Equation (2.19) can be rewritten as follows:

$$\begin{cases} \alpha_{i,1} = \dfrac{(X_3 - X_2)(Y_i - Y_2) - (Y_3 - Y_2)(X_i - X_2)}{(X_3 - X_2)(Y_1 - Y_2) - (Y_3 - Y_2)(X_1 - X_2)} \\[2ex] \alpha_{i,2} = \dfrac{(X_1 - X_3)(Y_i - Y_3) - (Y_1 - Y_3)(X_i - X_3)}{(X_1 - X_3)(Y_2 - Y_3) - (Y_1 - Y_3)(X_2 - X_3)} \\[2ex] \alpha_{i,3} = \dfrac{(X_2 - X_1)(Y_i - Y_1) - (Y_2 - Y_1)(X_i - X_1)}{(X_2 - X_1)(Y_3 - Y_1) - (Y_2 - Y_1)(X_3 - X_1)} \end{cases} . \tag{2.20}$$

Parameters $\alpha_{i,k} = c$ (c is a constant parameter.) in Eq. (2.20) define lines that are parallel to the sides of the leading triangle. Parallel lines associated with $\alpha_{i,k} = c$ are depicted in Fig. 2.3. For instance, $\alpha_{i,3} = c$ represents a line that is parallel to

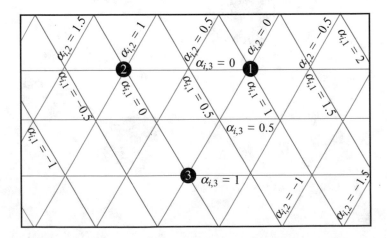

Fig. 2.3 Parameters $\alpha_{i,k}$=constant

the side s_{12}, i.e., s_{12} is the line segment connecting vertices 1 and 2 of the leading triangle. Notice that $\alpha_{i,k} = 1$ passes through the vertex k. Also, $\alpha_{i,k} = 0$ defines a line coinciding on the side of the leading triangle that is not passing through the vertex k. Moreover, $\alpha_{i,1}$, $\alpha_{i,2}$, and $\alpha_{i,3}$ are all positive, if the follower i is located inside the leading triangle.

Entries of the matrix $Q \in \mathbb{R}^{2 \times 2}(Q_{11}, Q_{12}, Q_{21}$, and $Q_{22})$ and the vector $D \in \mathbb{R}^2$ (D_1 and D_2) can be uniquely obtained based on X and Y components of the leaders' positions as follows:

$$\begin{bmatrix} Q_{11}(t) \\ Q_{12}(t) \\ Q_{21}(t) \\ Q_{22}(t) \\ D_1(t) \\ D_2(t) \end{bmatrix} = \begin{bmatrix} X_1 & Y_1 & 0 & 0 & 1 & 0 \\ X_2 & Y_2 & 0 & 0 & 1 & 0 \\ X_3 & Y_3 & 0 & 0 & 1 & 0 \\ 0 & 0 & X_1 & Y_1 & 0 & 1 \\ 0 & 0 & X_2 & Y_2 & 0 & 1 \\ 0 & 0 & X_3 & Y_3 & 0 & 1 \end{bmatrix}^{-1} \begin{bmatrix} x_1(t) \\ x_2(t) \\ x_3(t) \\ y_1(t) \\ y_2(t) \\ y_3(t) \end{bmatrix}. \tag{2.21}$$

Spatial Description of Homogeneous Deformation: Material descriptions of velocity and acceleration fields of a continuum, deforming under a homogeneous transformation, are given by

$$v = \dot{r} = \dot{Q}R + \dot{D} \tag{2.22}$$

$$a = \ddot{r} = \ddot{Q}R + \ddot{D}. \tag{2.23}$$

Let R in Eqs. (2.22) and (2.23) be replaced by $Q^{-1}(r - D)$, then

$$v = \dot{Q}Q^{-1}r + (\dot{D} - \dot{Q}Q^{-1}D) \tag{2.24}$$

$$a = \ddot{Q}Q^{-1}r + (\ddot{D} - \ddot{Q}Q^{-1}D) \tag{2.25}$$

are the spatial descriptions for the velocity and acceleration fields of homogeneous deformation. Components of $v(x_1, x_2, t) = v_1(x_1, x_2, t)\hat{\mathbf{e}}_1 + v_2(x_1, x_2, t)\hat{\mathbf{e}}_2$ and $a(x_1, x_2, t) = a_1(x_1, x_2, t)\hat{\mathbf{e}}_1 + a_2(x_1, x_2, t)\hat{\mathbf{e}}_2$ are obtained as

$$v_1(x_1, x_2, t) = c(t)x_1 + d(t)x_2 + e(t) \tag{2.26}$$

$$v_2(x_1, x_2, t) = f(t)x_1 + g(t)x_2 + h(t) \tag{2.27}$$

$$a_1(x_1, x_2, t)) = l(t)x_1 + m(t)x_2 + o(t) \tag{2.28}$$

$$a_2(x_1, x_2, t) = p(t)x_1 + q(t)x_2 + s(t) \tag{2.29}$$

where

$$c(t) = \frac{\dot{Q}_{11}Q_{22} - \dot{Q}_{12}Q_{21}}{|Q(t)|},$$

$$d(t) = \frac{\dot{Q}_{12}Q_{11} - \dot{Q}_{11}Q_{12}}{|Q(t)|},$$

$$f(t) = \frac{\dot{Q}_{21}Q_{22} - \dot{Q}_{22}Q_{21}}{|Q(t)|},$$

$$g(t) = \frac{\dot{Q}_{22}Q_{11} - \dot{Q}_{21}Q_{12}}{|Q(t)|},$$

$$e(t) = \dot{D}_1 - \frac{(\dot{Q}_{11}Q_{22} - \dot{Q}_{12}Q_{21})D_1 + (\dot{Q}_{12}Q_{11} - \dot{Q}_{11}Q_{12})D_2}{|Q(t)|},$$

$$h(t) = \dot{D}_2 - \frac{(\dot{Q}_{21}Q_{22} - \dot{Q}_{22}Q_{21})D_1 + (\dot{Q}_{22}Q_{11} - \dot{Q}_{21}Q_{12})D_2}{|Q(t)|},$$

$$l(t) = \frac{(\ddot{Q}_{11}Q_{22} - \ddot{Q}_{12}Q_{21})}{|Q(t)|},$$

$$m(t) = \frac{\ddot{Q}_{12}Q_{11} - \ddot{Q}_{11}Q_{12}}{|Q(t)|}$$

$$p(t) = \frac{\ddot{Q}_{21}Q_{22} - \ddot{Q}_{22}Q_{21}}{|Q(t)|}$$

$$q(t) = \frac{\ddot{Q}_{22}Q_{11} - \ddot{Q}_{21}Q_{12}}{|Q(t)|}$$

$$s(t) = \ddot{D}_2 - \frac{(\ddot{Q}_{21}Q_{22} - \ddot{Q}_{22}Q_{21})D_1 + (\ddot{Q}_{22}Q_{11} - \ddot{Q}_{21}Q_{12})D_2}{|Q(t)|}.$$

2.1.3 *Homogeneous Deformation of Agents with Finite Size*

Consider an MAS that consists of N agents and move collectively in $\mathbb{R}^n$ ($n = 1, 2, 3$). Each agent is considered as a ball with radius ε. Because homogeneous transformation can change interagent distance among agents, it is necessary to assure that no two agents collide when the MAS deforms homogeneously.

Assume that m_1 and m_2 are index numbers of two agents having closest distance in the MAS initial configuration. To assure collision avoidance, no two agents should not get closer than 2ε ($\eta \in \mathbb{R}_+$). This requirement is assured, if

$$\forall t \geq t_0 ||r_{m_1,HT} - r_{m_2,HT}|| > 2\varepsilon. \tag{2.30}$$

By considering definition of homogeneous deformation,

$$r_{m_1,HT} - r_{m_2,HT} = Q(R_{m_1} - R_{m_2}). \tag{2.31}$$

Therefore,

$$\left(r_{m_1,HT} - r_{m_2,HT}\right)^T \left(r_{m_1,HT} - r_{m_2,HT}\right) \geq \left(R_{m_1} - R_{m_2}\right)^T Q^T Q \left(R_{m_1} - R_{m_2}\right). \quad (2.32)$$

and

$$\left\|r_{m_1,HT} - r_{m_2,HT}\right\| \geq \lambda_{min}\left(\sqrt{Q^T Q}\right)\left\|R_{m_1} - R_{m_2}\right\|. \quad (2.33)$$

By using polar decomposition, the Jacobian matrix Q can be expressed as

$$Q = R_O U_D \quad (2.34)$$

where U_D is a symmetric pure deformation matrix and R_O is an orthogonal matrix (See Appendix A). Therefore, eigenvalues of the matrix

$$Q^T Q = U_D{}^T U_D \quad (2.35)$$

are all positive, and eigenvectors are mutually orthogonal.

Let

$$\lambda_{min} = \frac{2\varepsilon}{\|R_{m_1} - R_{m_2}\|} \quad (2.36)$$

be the lower bound for the eigenvalues of the matrix $\sqrt{Q^T(t)Q(t)}$ ($\forall t \geq t_0$), then the condition (2.30) is satisfied and no two agents get closer than 2ε ($\forall t \geq t_0, \|r_{m_1,HT}(t) - r_{m_2,HT}(t)\| \geq 2\varepsilon$).

Example 2.1. Consider an MAS with five agents (three leader and two followers) with initial and final formations shown in Fig. 2.4. Leaders are initially placed

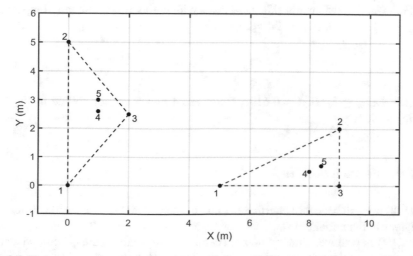

Fig. 2.4 Homogeneous transformation of agents in Example 2.2

at $(X_1, Y_1) = (0,0)$, $(X_2, Y_2) = (0,5)$, and $(X_3, Y_3) = (2, 2.5)$. They are ultimately stop at $(X_{F,1}, Y_{F,1}) = (5,0)$, $(X_{F,2}, Y_{F,2}) = (9,2)$, and $(X_{F,3}, Y_{F,3}) = (9,0)$. By using Eq. (2.21),

$$
\begin{bmatrix} Q_{11} \\ Q_{12} \\ Q_{21} \\ Q_{22} \\ D_1 \\ D_2 \end{bmatrix} = \begin{bmatrix} 0 & 0 & 0 & 0 & 1 & 0 \\ 0 & 5 & 0 & 0 & 1 & 0 \\ 2 & 2.5 & 0 & 0 & 1 & 0 \\ 0 & 0 & 0 & 0 & 0 & 1 \\ 0 & 0 & 0 & 5 & 0 & 1 \\ 0 & 0 & 2 & 2.5 & 0 & 1 \end{bmatrix}^{-1} \begin{bmatrix} 5 \\ 9 \\ 9 \\ 0 \\ 2 \\ 0 \end{bmatrix} = \begin{bmatrix} 1 \\ 0.8 \\ -0.5 \\ 0.4 \\ 5 \\ 0 \end{bmatrix}.
$$

Therefore,

$$
U_D^2 = Q^T Q = \begin{bmatrix} 1.2500 & 0.6000 \\ 0.6000 & 0.8000 \end{bmatrix},
$$

$\lambda_1(Q^T Q) = 0.3842$ and $\lambda_2(Q^T Q) = 1.6658$. Moreover, $\mathbf{n}_1 = 0.5696\hat{e}_x - 0.8219\hat{e}_y$, $\mathbf{n}_2 = -0.8219\hat{e}_x - 0.5696\hat{e}_y$ are the eigenvectors of both matrices $Q^T Q$ and $U_D = \sqrt{Q^T Q}$. Thus,

$$
U_D = \begin{bmatrix} \mathbf{n}_1 & \mathbf{n}_2 \end{bmatrix} \begin{bmatrix} \sqrt{0.3842} & 0 \\ 0 & \sqrt{1.6658} \end{bmatrix} \begin{bmatrix} \mathbf{n}_1 & \mathbf{n}_2 \end{bmatrix}^T = \begin{bmatrix} 1.0730 & 0.3141 \\ 0.3141 & 0.8375 \end{bmatrix}
$$

$\lambda_1(U_D) = 0.6198$ and $\lambda_2(Q^T Q) = 1.2907$. Given U_D and Q,

$$
R_O = Q U_D^{-1} = \begin{bmatrix} 0.7328 & 0.6805 \\ -0.6805 & 0.7328 \end{bmatrix}.
$$

Follower 4 and 5 are considered as disks with radius $10cm$ with initial separation $\|R_4 - R_5\| = 40cm$. Hence, $2\varepsilon = 20cm$ and

$$
\lambda_{min} = \frac{20}{40} = 0.5.
$$

Because $\lambda_{min} \le \lambda_1(U_D) = 0.6198$, followers 4 and 5 do not collide in the final configuration.

2.1.4 Force Analysis

Let followers all have the same mass m, and they are all distributed inside the leading polytope at the initial time t_0. Then, the parameter $\alpha_{i,k}$ ($k = 1, 2, \ldots, n+1, i = n+2, \ldots, N$) is positive. Under a homogeneous transformation, the qth component of acceleration of the follower agent i ($i = n+2, \ldots, N$) becomes

$$\ddot{x}_{q,i}(t) = \sum_{k=1}^{n+1} \alpha_{i,k} \ddot{x}_{q,k}(t). \tag{2.37}$$

If $a_q(t)$ denotes the maximum for the magnitude of the qth component of the leaders' acceleration at the time t, i.e.

$$||\ddot{x}_{q,k}(t)|| \le a_q(t), \ k = 1, 2, \dots, n+1, \tag{2.38}$$

then the qth component of the acceleration of the follower i satisfies the following inequality:

$$||\ddot{x}_{q,i}(t)|| = ||\sum_{k=1}^{n+1} \alpha_{i,k} \ddot{x}_{q,k}(t)|| \le a_q(t) \sum_{k=1}^{n+1} \alpha_{i,k} = a_q(t). \tag{2.39}$$

Equation (2.39) yields an upper bound for maximum force required for the motion of the follower i. Let

$$f_i(t) = m \sum_{q=1}^{n} \ddot{x}_{q,k}(t) \hat{e}_q \tag{2.40}$$

be the force required for the motion of follower i at time t, then

$$||f_i(t)|| = m \sqrt{\sum_{q=1}^{n} \ddot{x}_{q,i}^2(t)} \le m \sqrt{\sum_{q=1}^{n} a_q^2(t)}. \tag{2.41}$$

Example 2.2. Consider an MAS with the initial formation shown in Fig. 2.5. As it is seen, 3 leaders are placed at the vertices of the leading triangle and 17 followers are distributed inside the leading triangle. Therefore, the parameter $\alpha_{i,k}$ ($i = 4, 5, \dots, 20$ and $k = 1, 2, 3$) is positive. In Fig. 2.6, paths of the leaders in the $X - Y$ plane are shown. Leaders initiate their motion from rest at $t = 0s$ and settle at $T = 60s$. Configurations of the leading triangle are depicted at the initial and final times, where both have the same area of $50m^2$.

Each follower has the mass $m = 1kg$. In Fig. 2.7, magnitudes of forces acting on the followers are shown by black curves. Also, the upper limit for these acting forces is obtained by Eq. (2.41) and shown by the red curve. As it is seen in Fig. 2.7, magnitudes of the acting forces do not exceed the upper limit $\sqrt{a_1^2 + a_2^2}$, where $a_1(t)$ and $a_2(t)$ are the supremum for the X and Y components of the leaders' accelerations over $t \in [0, 60]$, respectively.

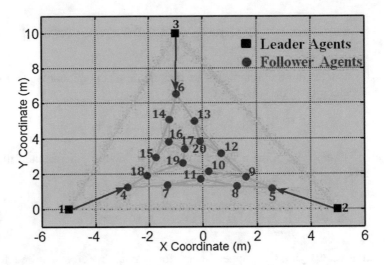

Fig. 2.5 Initial distribution of the agents in Example 2.2

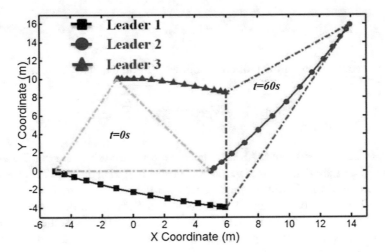

Fig. 2.6 Paths of the leaders

2.1.5 Path Planning: MAS As Particles of Newtonian Viscous Flow

A continuum can be considered as a portion of the incompressible Newtonian viscous flow, if the deformation mapping satisfies both the continuity and Navier-Stokes constitutive equations. Let $v(x_1, x_2, x_3, t) = v_1(x_1, x_2, x_3, t)\hat{e}_1 + v_2(x_1, x_2, x_3, t)\hat{e}_2 + v_3(x_1, x_2, x_3, t)\hat{e}_3$ be the spatial description of the velocity field of a continuum, then continuity equation is expressed by

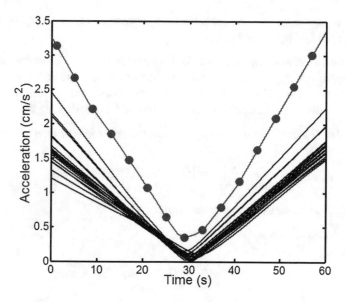

Fig. 2.7 Acting forces required for followers to deform as homogeneous transformation

$$\frac{\partial v_1}{\partial x_1} + \frac{\partial v_2}{\partial x_2} + \frac{\partial v_3}{\partial x_3} = 0. \qquad (2.42)$$

Also, the Navier-Stokes equations are expressed as follows:

$$\rho a_1 = -\frac{\partial p}{\partial x_1} + \mu \left(\frac{\partial^2 v_1}{\partial x_1^2} + \frac{\partial^2 v_1}{\partial x_2^2} + \frac{\partial^2 v_1}{\partial x_3^2} \right) \qquad (2.43)$$

$$\rho a_2 = -\frac{\partial p}{\partial x_2} + \mu \left(\frac{\partial^2 v_2}{\partial x_1^2} + \frac{\partial^2 v_2}{\partial x_2^2} + \frac{\partial^2 v_2}{\partial x_3^2} \right) \qquad (2.44)$$

$$\rho a_3 = -\frac{\partial p}{\partial x_3} + \mu \left(\frac{\partial^2 v_3}{\partial x_1^2} + \frac{\partial^2 v_3}{\partial x_2^2} + \frac{\partial^2 v_3}{\partial x_3^2} \right). \qquad (2.45)$$

It is noted that p is the pressure filed, ρ and μ are density and viscosity of the fluid flow, respectively. Additionally,

$$a_1(x_1, x_2, x_3, t) = \frac{\partial v_1}{\partial t} + v_1 \frac{\partial v_1}{\partial x_1} + v_2 \frac{\partial v_1}{\partial x_2} + v_3 \frac{\partial v_1}{\partial x_3} \qquad (2.46)$$

$$a_2(x_1, x_2, x_3, t) = \frac{\partial v_2}{\partial t} + v_1 \frac{\partial v_2}{\partial x_1} + v_2 \frac{\partial v_2}{\partial x_2} + v_3 \frac{\partial v_2}{\partial x_3} \qquad (2.47)$$

$$a_3(x_1, x_2, x_3, t) = \frac{\partial v_3}{\partial t} + v_1 \frac{\partial v_3}{\partial x_1} + v_2 \frac{\partial v_3}{\partial x_2} + v_3 \frac{\partial v_3}{\partial x_3} \qquad (2.48)$$

are the components of the acceleration $a(x_1,x_2,x_3,t) = a_1(x_1,x_2,x_3,t)\hat{e}_1 + a_2(x_1,x_2,x_3,t)\hat{e}_2 + a_3(x_1,x_2,x_3,t)\hat{e}_3$.

Holonomic Constraints Imposed on the Leaders' Motion: Continuity and Navier-Stokes equations impose two holonomic constraints on the leaders' motion, if homogeneous deformation of the MAS is treated as incompressible fluid flow. Before these two holonomic constraints are obtained, the following lemma is stated:

Lemma. *Let $Q \in \mathbb{R}^{n \times n}$ be a non-singular second order tensor, then*

$$\frac{d|Q|}{dt} = |Q|trace(\dot{Q}Q^{-1}). \tag{2.49}$$

Note that $|Q|$ is the determinant of the matrix Q.

Proof. It is known that

$$\frac{\partial |Q|}{\partial Q_{ij}} = Q_{ij}^{C} = M_{ji} \tag{2.50}$$

where Q_{ij}^{C} is the cofactor of the entry ij of the matrix Q, and

$$Q^{-1} = \frac{M^T}{|Q|}. \tag{2.51}$$

Thus,

$$\frac{d|Q|}{dt} = \sum_{i=1}^{n}\sum_{j=1}^{n} \frac{\partial |Q|}{\partial Q_{ij}}\dot{Q}_{ij} = \sum_{i=1}^{n}\sum_{j=1}^{n} M_{ji}\dot{Q}_{ij} = |Q|trace(Q\dot{Q}^{-1}). \tag{2.52}$$

Theorem 2.1. *Homogeneous transformation of a continuum can be considered as deformation of incompressible Newtonian viscous fluid flow, if*

$$\forall t \geq t_0, |Q(t)| = 1 \tag{2.53}$$

$$curl\ a = \mathbf{0} \tag{2.54}$$

Proof. Let the velocity of a homogeneous deformation satisfy the continuity Eq. (2.42), then

$$\nabla.v = trace(\dot{Q}Q^{-1}) = 0. \tag{2.55}$$

Therefore, the right-hand side of Eq. (2.52) vanishes and $|Q(t)|$ remains constant. Because $Q(t_0) = I_n \in \mathbb{R}^{n \times n}$, $|Q(t)| = 1$ at any time during deformation.

Under homogeneous deformation,

$$\nabla^2 v_i = 0,\ i = 1,2,3,$$

therefore, the Navier-Stokes equations simplify to

$$\rho a = -\nabla p. \tag{2.56}$$

Note that pressure field is smooth, therefore,

$$\frac{\partial^2 p}{\partial x_1 \partial x_2} = \frac{\partial^2 p}{\partial x_2 \partial x_1}$$

$$\frac{\partial^2 p}{\partial x_2 \partial x_3} = \frac{\partial^2 p}{\partial x_3 \partial x_2}$$

$$\frac{\partial^2 p}{\partial x_3 \partial x_1} = \frac{\partial^2 p}{\partial x_1 \partial x_3}$$

and *curl a* = **0**.

Corollary 1. *If homogeneous deformation of a continuum in* $\mathbb{R}^2$ *satisfies the continuity equation, then the area of the leading triangle is preserved during deformation. Therefore, the following holonomic constraint can be considered for motion of the leaders:*

$$C_1 : \frac{1}{2} \begin{vmatrix} x_1(t) & x_2(t) & x_3(t) \\ y_1(t) & y_2(t) & y_3(t) \\ 1 & 1 & 1 \end{vmatrix} = \tag{2.57}$$

$$x_1(y_2 - y_3) + x_2(y_3 - y_1) + x_3(y_1 - y_2) - 2a_0 = 0.$$

Note that

$$a_0 = \frac{1}{2} \begin{vmatrix} X_1 & X_2 & X_3 \\ Y_1 & Y_2 & Y_3 \\ 1 & 1 & 1 \end{vmatrix}. \tag{2.58}$$

is the area of the leading triangle at the initial time t_0.

Corollary 2. *If homogeneous deformation satisfies the Navier-Stokes equations, then motion of the leaders should satisfy the following constraint:*

$$C_2 = \ddot{x}_1(x_2 - x_3) + \ddot{x}_2(x_3 - x_1) + \ddot{x}_3(x_1 - x_2) + \ddot{y}_1(y_3 - y_2) + \ddot{y}_2(y_1 - y_3) + \ddot{y}_3(y_2 - y_1) = 0. \tag{2.59}$$

Minimum Acceleration Norm of Homogeneous Deformation: In this section, optimal trajectories of the leaders minimizing acceleration norm of homogeneous transformation are determined, where leaders' positions satisfy the holonomic constraints (2.57) and (2.59) at any time during MAS evolution, and initial and final positions and velocities of leaders are known. It is assumed that leaders' positions are updated by

$$\ddot{x}_i = q_i, \ i = 1, \ 2, \ 3, \tag{2.60}$$

$$\ddot{y}_i = q_i, \ i = 4, \ 5, \ 6. \tag{2.61}$$

The dynamics of the leaders can be converted into a state space representation,

$$\dot{P} = A_p P + B_p U_p, \tag{2.62}$$

where $P = [p_i] \in \mathbb{R}^{12}, p_1 = x_1, p_2 = x_2, p_3 = x_3, p_4 = y_1, p_5 = y_2, p_6 = y_3, p_7 = \dot{x}_1,$
$p_8 = \dot{x}_2, p_9 = \dot{x}_3, p_{10} = \dot{y}_1, p_{11} = \dot{y}_2, p_{12} = \dot{y}_3,$

$$U_p = \begin{bmatrix} q_1 & \cdots & q_6 \end{bmatrix}^T$$

$$A_p = \begin{bmatrix} \mathbf{0} & I_6 \\ \mathbf{0} & \mathbf{0} \end{bmatrix}^T$$

$$B_p = \begin{bmatrix} \mathbf{0} \\ I_6 \end{bmatrix}^T.$$

It is noted that $\mathbf{0}, I_6 \in \mathbb{R}^{6 \times 6}$ are zero-entry and identity matrices, respectively.
   The constraints (2.57) and (2.59) can be rewritten as

$$C_1' : \ q_1(p_5 - p_6) + q_2(p_6 - p_4) + q_3(p_4 - p_5) + q_4(p_3 - p_2) +$$
$$q_5(p_1 - p_3) + q_6(p_2 - p_1) + p_7(p_{11} - p_{12}) + p_8(p_{12} - p_{10}) + \tag{2.63}$$
$$p_9(p_{10} - p_{11}) + p_{10}(p_9 - p_8) + p_{11}(p_7 - p_9) + p_{12}(p_8 - p_7) = 0.$$

$$C_2 : \ q_1(p_2 - p_3) + q_2(p_3 - p_1) + q_3(p_1 - p_2) + q_4(p_5 - p_6)$$
$$+ q_6(p_6 - p_4) + q_6(p_4 - p_5) = 0. \tag{2.64}$$

It is noted that the constraint Eq. (2.63) is obtained by taking the second time
derivative from Eq. (2.57).
   Leaders start their motion from the rest at the initial time $t = 0$ and they stop in a
finite horizon of time T, where the leaders' initial and final positions are known.

Objective Function: The objective is to minimize the acceleration norm of homogeneous transformation of the MAS, where leaders satisfy the constraints C_1' and C_2
(Eqs. (2.63) and (2.64)). Therefore,

$$J = \int_0^T \{ U_p^T U_p + \lambda^T (A_p P + B_p U_p - \dot{P}) + \gamma_1 C_1' + \gamma_2 C_2 \} dt \tag{2.65}$$

is the cost function that is desired to be minimized, where $\lambda \in \mathbb{R}^{12}$ is the costate
vector, γ_1 and γ_2 are Lagrange multipliers. This is a well-known two point boundary

value problem that can be solved by using calculus of variation. The cost J is minimized, if

$$\delta J = \int_0^T \left\{ \sum_{i=1}^6 (2q_i \delta q_i + \delta \lambda_i (p_{i+6} - \dot{p}_i) + \right.$$

$$\lambda_i (\delta p_{i+6} - \delta \dot{p}_i) + \delta \lambda_{i+6} (q_i - \dot{p}_{i+6}) + \tag{2.66}$$

$$\left. \lambda_{i+6} (\delta q_i - \delta \dot{p}_{i+6})) + \delta \gamma_1 C_1' + \gamma_1 \delta C_1' + \delta \gamma_2 C_2 + \gamma_2 \delta C_2 \right\} dt = 0.$$

Therefore, X and Y components of the leaders' positions are determined by solving the following 24th order dynamics:

$$\dot{S} = A_s S, \tag{2.67}$$

where

$$S = \begin{bmatrix} P \\ \lambda \end{bmatrix} \tag{2.68}$$

$$A_s = \begin{bmatrix} \mathbf{0} & I_6 & \mathbf{0} & \mathbf{0} \\ -\frac{1}{2}(\gamma_1 K_1 + \gamma_2 K_3) & \mathbf{0} & \mathbf{0} & -\frac{1}{2}I_6 \\ \frac{1}{2}(\gamma_1^2 + \gamma_2^2)K_2 & \mathbf{0} & \mathbf{0} & \frac{1}{2}(\gamma_1 K_1 - \gamma_2 K_3) \\ \mathbf{0} & -2\gamma_1 K_1 & -I_6 & \mathbf{0} \end{bmatrix} \tag{2.69}$$

$$K_1 = \begin{bmatrix} 0 & 0 & 0 & 0 & 1 & -1 \\ 0 & 0 & 0 & -1 & 0 & 1 \\ 0 & 0 & 0 & 1 & -1 & 0 \\ 0 & -1 & 1 & 0 & 0 & 0 \\ 1 & 0 & -1 & 0 & 0 & 0 \\ -1 & 1 & 0 & 0 & 0 & 0 \end{bmatrix} \tag{2.70}$$

$$K_2 = \begin{bmatrix} 2 & -1 & -1 & 0 & 0 & 0 \\ -1 & 2 & -1 & 0 & 0 & 0 \\ -1 & -1 & 2 & 0 & 0 & 0 \\ 0 & 0 & 0 & 2 & -1 & -1 \\ 0 & 0 & 0 & -1 & 2 & -1 \\ 0 & 0 & 0 & -1 & -1 & 2 \end{bmatrix} \tag{2.71}$$

$$
K_3 = \begin{bmatrix} 0 & 1 & -1 & 0 & 0 & 0 \\ -1 & 0 & 1 & 0 & 0 & 0 \\ 1 & -1 & 0 & 0 & 0 & 0 \\ 0 & 0 & 0 & 0 & 1 & -1 \\ 0 & 0 & 0 & -1 & 0 & 1 \\ 0 & 0 & 0 & 1 & -1 & 0 \end{bmatrix}.
\tag{2.72}
$$

It is noted that γ_1 and γ_2 in Eq. (2.69) are equal to

$$
\gamma_1 = \frac{2\tau - \sigma}{\rho}
\tag{2.73}
$$

$$
\gamma_2 = -\frac{\varphi}{\rho},
\tag{2.74}
$$

where

$$
\rho = (p_5 - p_6)^2 + (p_6 - p_4)^2 + (p_4 - p_5)^2 + (p_3 - p_2)^2 + (p_1 - p_3)^2 + (p_2 - p_1)^2
\tag{2.75}
$$

$$
\tau = p_7(p_{11} - p_{12}) + p_8(p_{12} - p_{10}) + p_9(p_{10} - p_{11}) + p_{10}(p_9 - p_8) + p_{11}(p_7 - p_9) +
$$
$$
p_{12}(p_8 - p_7)
\tag{2.76}
$$

$$
\sigma = \lambda_7(p_5 - p_6) + \lambda_8(p_6 - p_4) + \lambda_9(p_4 - p_5) + \lambda_{10}(p_3 - p_2) + \lambda_{11}(p_1 - p_3) +
$$
$$
\lambda_{12}(p_2 - p_1)
\tag{2.77}
$$

$$
\varphi = \lambda_7(p_3 - p_2) + \lambda_8(p_1 - p_3) + \lambda_9(p_2 - p_1) + \lambda_{10}(p_6 - p_5) + \lambda_{11}(p_4 - p_6) +
$$
$$
\lambda_{12}(p_5 - p_4).
\tag{2.78}
$$

Furthermore, optimal control inputs become

$$
q_1 = -\frac{1}{2}(\lambda_7 + \gamma_1(p_5 - p_6) + \gamma_2(p_2 - p_3)),
\tag{2.79}
$$

$$
q_2 = -\frac{1}{2}(\lambda_8 + \gamma_1(p_6 - p_4) + \gamma_2(p_3 - p_1)),
\tag{2.80}
$$

$$
q_3 = -\frac{1}{2}\lambda_9 + \gamma_1(p_4 - p_5) + \gamma_2(p_1 - p_2)),
\tag{2.81}
$$

$$
q_4 = -\frac{1}{2}(\lambda_{10} + \gamma_1(p_3 - p_2) + \gamma_2(p_5 - p_6)),
\tag{2.82}
$$

$$q_5 = -\frac{1}{2}(\lambda_{11} + \gamma_1(p_1 - p_3) + \gamma_2(p_6 - p_4)), \tag{2.83}$$

$$q_6 = -\frac{1}{2}(\lambda_{12} + \gamma_1(p_5 - p_6) + \gamma_2(p_4 - p_5)). \tag{2.84}$$

Numerical Solution: It is difficult to obtain an analytic solution for the optimal trajectories minimizing acceleration norm of MAS evolution. This is because the dynamics of Eq. (2.67) is nonlinear and 24th order, where the initial values for the co-estate vector λ are not defined. Therefore, trajectories of the leaders are found by using a distributed gradient algorithm.

First, $\gamma_1(t)$ and $\gamma_2(t)$ are estimated by $\gamma_{11}(t)$ and $\gamma_{21}(t)$, over the time interval $t \in [t_0, T]$. Then $\gamma_1(t)$ and $\gamma_2(t)$ are kept updating until the solution of the optimal trajectories are obtained.

Let $\Phi_k(t, t_0)$, P_k, and λ_k denote state transition matrix, control state, and costate at the attempt ($k = 1, 2, 3, \ldots$), then

$$\forall t \geq t_0, \begin{bmatrix} P_k(t) \\ \lambda_k(t) \end{bmatrix} = \phi_k(t, t_0) \begin{bmatrix} P_k(t_0) \\ \lambda_k(t_0) \end{bmatrix} = \begin{bmatrix} \phi_{11k}(t, t_0) & \phi_{12k}(t, t_0) \\ \phi_{21k}(t, t_0) & \phi_{22k}(t, t_0) \end{bmatrix} \begin{bmatrix} P(t_0) \\ \lambda(t_0) \end{bmatrix}. \tag{2.85}$$

Thus,

$$P(T) = P_k(T) = \phi_{11k}(T, t_0)P(t_0) + \phi_{12k}\lambda_k(t_0). \tag{2.86}$$

Notice that $P(t_0)$ and $P(T)$ are both known because positions and velocities of the leaders are given at the initial time t_0 and final time T. Hence, $\lambda_k(t_0)$ is obtained as follows:

$$\lambda_k(t_0) = \phi_{12k}(T, t_0)^{-1}(P(T) - \phi_{11k}(T, t_0)P(t_0)). \tag{2.87}$$

Given $P(t_0)$ and $\lambda_k(t_0)$, $P_k(t)$ and $\lambda_k(t)$ (the kth estimations of $P(t)$ and $\lambda(t)$) can be updated by Eq. (2.85). Then, $\gamma_{1k}(t)$ and $\gamma_{2k}(t)$ can be updated by Eqs. (2.73) and (2.74). This process is continued until

$$\forall t \in [t_0, T], \; |\gamma_{ik}(t) - \gamma_{ik-1}(t)| \to 0, \; i = 1, 2. \tag{2.88}$$

Example 2.3. Suppose that leaders are initially placed at $(-5, 0)$, $(5, 0)$, and $(-1, 10)$. They start their motion from rest at $t_0 = 0s$ and ultimately stop at $(7, -5)$, $(15, 15)$, and $(7, 7.5)$ at $T = 60s$. Leaders are restricted to satisfy the holonomic constraints C_1' and C_2 (Eqs. (2.63) and (2.64)) at any time $t \in [0, 60]s$. The optimal paths of the leaders minimizing the objective function (2.65) are shown in Fig. 2.8. Also, the optimal control inputs q_i are shown versus time in Fig. 2.9.

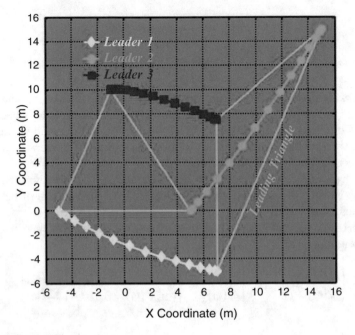

Fig. 2.8 Optimal paths of the leader agents

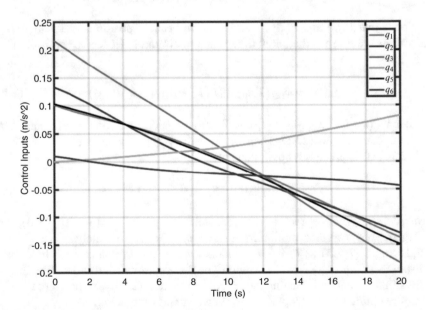

Fig. 2.9 Optimal control inputs

2.2 Evolution of a Multi-Agent System with Constrained Dynamics

Suppose that position of the follower $i \in V_F$ is updated by

$$\begin{cases} \dot{Z}_i = f(Z_i, u_i) \\ r_i = h(Z_i) \end{cases} \tag{2.89}$$

where the nonlinear dynamics (2.89) is both state controllable and state observable, $Z_i \in \mathbb{R}^{n_z}$ is the state vector, $u_i \in \mathbb{R}^{n_u}$ is the control input that belongs to a connected set $U_i \subset \mathbb{R}^{n_u}$, and actual position of the follower i, $r_i \in \mathbb{R}^n$, is considered as the control output.

Let $\Phi_i^k \subset \mathbb{R}^n$ be considered as the reachable displacement set for the follower i at the $t = t_{k+1}$. Then, Φ_i^k can be reached by the follower i at the time $t = t_{k+1}$, if the follower i chooses an admissible control $u_i \in U_i$ during $t \in [t_k, t_{k+1}]$. It is noted that the set $\Phi_i^k \subset \mathbb{R}^n$ is connected because the system is controllable. Also, Φ_i^k includes the origin since displacement of the follower i is $0 \in \mathbb{R}^n$ at the time t_k.

It is desired that

- MAS configuration at the time $t = t_{k+1}$ is a homogeneous deformation of the MAS formation at t_k, and
- $\forall t \in [t_k, t_{k+1}], \forall i \in V_F, ||r_i(t) - r_i(t_k)|| \in \Phi_i^k$.

In the theorem below, it is shown under what conditions the above objectives can be achieved.

Theorem 2.2. *Let*

(i) Φ_i^k *be the reachable displacement set of the follower i ($\forall i \in V_F$), where each follower is permitted to choose an admissible control input $u_i \in U_i$ during the time interval $[t_k, t_{k+1}]$,*

(ii) $\Phi_k = \bigcap_{i \in V_F} \Phi_i^k$, *and*

(iii) $\forall t \in [t_k, t_{k+1}], \forall i \in V_L, ||r_i(t) - r_i(t_k)|| \in \Omega_r^{k+1}$,

where

$$\Omega_r^{k+1} = \{ d \in \Phi_k | \ ||d|| \leq r^{k+1} \}$$

is the biggest ball inside Φ_k with radius r^{k+1}. Then, followers can reach the desired positions defined by a homogeneous deformation at t_{k+1} ($k = 1, 2, \ldots$).

Proof. Let

$$\forall j \in V_L, \ d_j^{k+1} = r_j^{k+1} - r_j^k \in \Omega_r^{k+1},$$

denote displacement of the leader j, then desired displacement of the follower i, issued by a homogeneous deformation, is obtained as follows:

$$d_i^{k+1} = \sum_{j=1}^{n+1} \alpha_{i,j}(r_j^{k+1} - r_j^k) = \sum_{j=1}^{n+1} \alpha_{i,j} d_j^{k+1}. \qquad (2.90)$$

Let

$$r^{k+1} = max\{d_1^{k+1}, d_2^{k+1}, \ldots, d_{n+1}^{k+1}\}$$

assign a maximum for leaders' displacements at t_{k+1}, then

$$||d_i^{k+1}|| = ||\sum_{j=1}^{n+1} \alpha_{i,j}(r_j^{k+1} - r_j^k)|| \leq \sum_{j=1}^{n+1} \alpha_{i,j}||d_j^{k+1}|| \leq r^{k+1}, \forall i \in V_F. \qquad (2.91)$$

Therefore,

$$\forall i \in V_F, \ ||d_i^{k+1}|| \in \Omega_r^{k+1} \subset \Phi_i^k \subset \Phi_k.$$

This implies that the desired position issued by a homogeneous deformation at t_{k+1} can be reached by the follower $i \in V_F$.

Desired Positions of the Followers: It is assumed that each leader j ($j = 1, 2, \ldots, n+1$) moves on a line segment given by

$$r_j(t) = \frac{r_j^{k+1} - r_j^k}{t_{k+1} - t_k}(t - t_k) + r_j^k \qquad (2.92)$$

where

$$||r_j^{k+1} - r_j^k|| \leq r^{k+1}. \qquad (2.93)$$

Consequently, followers acquire homogeneous deformation under no interagent communication by knowing only positions of the leaders at distinct times t_0, t_1, $\ldots$, t_f. Desired position of the follower i, defined by a homogeneous deformation, becomes

$$\forall t \in [t_k, t_{k+1}], \ r_{iHT}^k(t) = \sum_{j=1}^{n+1} \alpha_{i,j} \frac{r_j^{k+1} - r_j^k}{t_{k+1} - t_k}(t - t_k) + r_j^k. \qquad (2.94)$$

Homogeneous Deformation of Wheeled Robots: Let each follower i be a wheeled robot with the dynamics

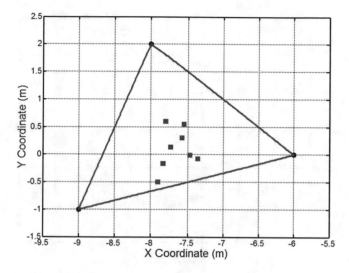

Fig. 2.10 Initial distribution of the agents in Example 2.4

Table 2.1 Initial heading angles of follower wheeled robots

Heading Angle	4	5	6	7	8	9	10	11
θ_i	45	30	87	123	135	130	65	100

$$\begin{cases} \dot{x}_i = u_i \cos \theta_i \\ \dot{y}_i = u_i \sin \theta_i \\ \dot{\theta}_i = \omega_i \end{cases}, \tag{2.95}$$

where the magnitude of u_i and ω_i are limited by

$$\begin{cases} |u_i| \le 1 \\ |\omega_i| \le 1 \end{cases}. \tag{2.96}$$

For simulation, evolution of an MAS containing 11 agents (3 leaders and 8 follower unicycle robots) is considered. The MAS has the initial formation that is shown in Fig. 2.10. Also, initial followers' orientations are listed in Table 2.1.

In Fig. 2.11, reachable displacement set $\Phi_5{}^0$ of the follower agent 5 at $t_1 = \pi s$ is illustrated. All reachable displacement sets ($\Phi_i{}^0, i = 4, 5, \ldots, 11$) are also illustrated in Fig. 2.12. As it is observed, the disk Ω_0 has the radius $2m$ and it is centered at the origin. Notice that Ω_0 is a subset of

$$\Phi_0 = \bigcap_{i=4}^{11} \Phi_i{}^0.$$

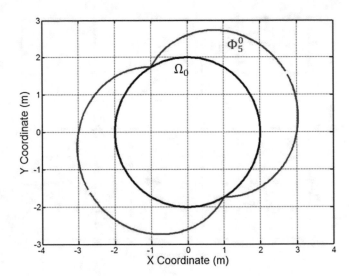

Fig. 2.11 Reachable displacement set $\Phi_5{}^0$

Table 2.2 Parameters $\alpha_{i,j}$ of the follower i with initial position shown in Fig. 2.10

	$X_i(m)$	$Y_i(m)$	$\alpha_{i,1}$	$\alpha_{i,2}$	$\alpha_{i,3}$
i=1	−9.0000	−1.0000	–	–	–
i=2	−6.0000	0	–	–	–
i=3	−8.0000	2.0000	–	–	–
i=4	−7.9000	−0.5000	0.6000	0.3500	0.0500
i=5	−7.8000	0.6000	0.3000	0.2500	0.4500
i=6	−7.7333	0.1333	0.4000	0.3333	0.2667
i=7	−7.5500	0.5500	0.2500	0.3500	0.4000
i=8	−7.5800	0.3000	0.3200	0.3700	0.3100
i=9	−7.3600	−0.0800	0.3600	0.5000	0.1400
i=10	−7.8333	−0.1667	0.5000	0.3333	0.1667
i=11	−7.4700	−0.0100	0.3700	0.4500	0.1800

Therefore, leaders are not permitted to transverse more than $2m$ during the time interval $t \in [0, \pi]$ seconds. By using Eq. (2.94), X and Y components of desired position of the follower $i \in V_F$ are obtained as follows:

$$\begin{cases} x_{i,HT}{}^k = \dfrac{t - t_k}{t_{k+1} - t_k} \sum_{j=1}^{n+1} \alpha_{i,j}(x_j{}^{k+1} - x_j{}^k) + x_j{}^k \\ y_{i,HT}{}^k = \dfrac{t - t_k}{t_{k+1} - t_k} \sum_{j=1}^{n+1} \alpha_{i,j}(y_j{}^{k+1} - y_j{}^k) + y_j{}^k \end{cases}. \tag{2.97}$$

In Eq. (2.97) t denotes time, $r_j{}^k = x_j{}^k \hat{\mathbf{e}}_x + y_j{}^k \hat{\mathbf{e}}_y$ and $r_j{}^{k+1} = x_j{}^{k+1} \hat{\mathbf{e}}_x + y_j{}^{k+1} \hat{\mathbf{e}}_y$. Initial centroid positions of the follower unicycles and the corresponding parameters $\alpha_{i,j}$, specified by Eq. (2.18), are listed in Table 2.2.

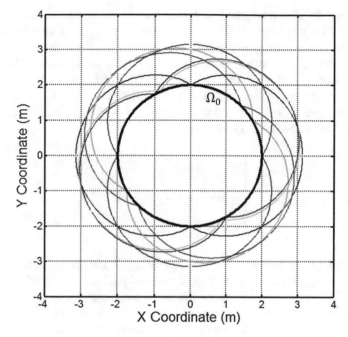

Fig. 2.12 The biggest disk Ω_0 inside the allowable displacement region

By taking second time derivative from Eq. (2.95), dynamics of evolution of the follower unicycle i becomes

$$\begin{bmatrix} \ddot{x}_i \\ \ddot{y}_i \end{bmatrix} = \begin{bmatrix} \cos\theta_i & -u_i\sin\theta_i \\ \sin\theta_i & u_i\cos\theta_i \end{bmatrix} \begin{bmatrix} \dot{u}_i \\ \dot{\omega}_i \end{bmatrix} + \begin{bmatrix} -2\dot{u}_i\omega_i\sin\theta_i - u_i\omega_i^2\cos\theta_i \\ 2\dot{u}_i\omega_i\cos\theta_i - u_i\omega_i^2\sin\theta_i \end{bmatrix}. \tag{2.98}$$

If the linear velocity u_i does not vanish at any time t, then $\ddot{u}_i$ and $\dot{\omega}_i$ can be related to the centroid jerk (time derivative of the centroid acceleration) of the follower i by

$$\begin{bmatrix} \ddot{u}_i \\ \dot{\omega}_i \end{bmatrix} = \begin{bmatrix} \cos\theta_i & -u_i\sin\theta_i \\ \sin\theta_i & u_i\cos\theta_i \end{bmatrix}^{-1} \begin{bmatrix} \dddot{x}_i + 2\dot{u}_i\omega_i\sin\theta_i + u_i\omega_i^2\cos\theta_i \\ \dddot{y}_i - 2\dot{u}_i\omega_i\cos\theta_i + u_i\omega_i^2\sin\theta_i \end{bmatrix}. \tag{2.99}$$

Let

$$\begin{bmatrix} \dddot{x}_i \\ \dddot{y}_i \end{bmatrix} = \begin{bmatrix} \dddot{x}_{i,HT} \\ \dddot{y}_{i,HT} \end{bmatrix} + \zeta_i \begin{bmatrix} \ddot{x}_{i,HT} - \ddot{x}_i \\ \ddot{y}_{i,HT} - \ddot{y}_i \end{bmatrix} + \gamma_i \begin{bmatrix} \dot{x}_{i,HT} - \dot{x}_i \\ \dot{y}_{i,HT} - \dot{y}_i \end{bmatrix} + \eta_i \begin{bmatrix} x_{i,HT} - x_i \\ y_{i,HT} - y_i \end{bmatrix}, \tag{2.100}$$

then r_i asymptotically converges to $r_{i,HT}$ when $\zeta_i > 0$, $\gamma_i > 0$, $\eta_i > 0$ and $\gamma_i\zeta_i > \eta_i$. This is because the transient error (the difference between r_i and $r_{i,HT}$) converges to zero as $t \to \infty$. It is noted that $x_{i,HT}$ and $y_{i,HT}$ are the X and Y components of the desired centroid position of the robot $i \in V_F$ that is defined by a homogeneous transformation.

Discussion: As aforementioned, if leader agents do not leave the disk Ω_0 in π s, then followers can reach the desired positions prescribed by a homogeneous deformation at the time $t_1 = t_0 + \pi$. One possibility for a follower i is to move along the circular trajectory, connecting its $r_{i,HT}{}^0$ (the desired centroid position at the initial time t_0) and $r_{i,HT}{}^1$ (the desired centroid position at the time t_1), through choosing constant linear and angular velocities. For example, consider the follower 8 whose centroid is initially positioned at $r_{8,HT}{}^0 = -7.58\hat{e}_x + 0.30\hat{e}_y$. Then, it can reach $r_{8,HT}{}^1 = -7.58\hat{e}_x + 2.30\hat{e}_x$ by choosing

$$
\begin{cases}
\dot{x}_i = \dfrac{1}{\sqrt{2}} \cos\left(\dfrac{-t}{2} + \dfrac{3\pi}{4}\right) \\[2mm]
\dot{y}_i = \dfrac{1}{\sqrt{2}} \sin\left(\dfrac{-t}{2} + \dfrac{3\pi}{4}\right) \\[2mm]
\dot{\theta}_i = -\dfrac{1}{2}
\end{cases}
\tag{2.101}
$$

In Fig. 2.13, the circular trajectory given by Eq. (2.101) is depicted by a continuous curve, where it connects the desired centroid positions of the follower 8 at the time $t_0 = 0s$ and $t_1 = \pi s$. As seen in Fig. 2.13, when circular trajectory is chosen, the follower 8 deviates from the desired trajectory that is shown by the dotted curve. Therefore, follower agents do not deform as homogeneous transformation during the time $t \in (0, \pi)$ and consequently interagent collision is not necessarily avoided.

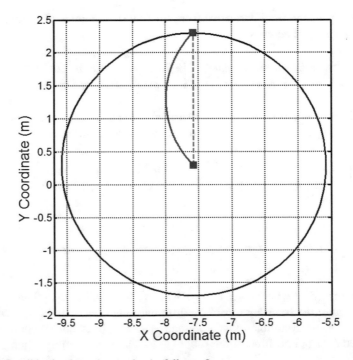

Fig. 2.13 Possible circular trajectory for the follower 8

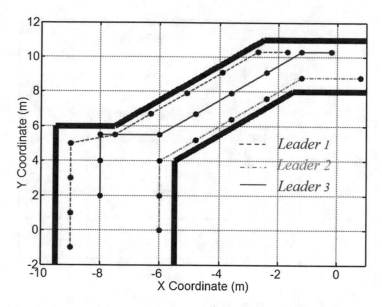

Fig. 2.14 Leaders' paths inside the narrow channel in Example 2.4

Notice that followers can fully track the desired line segment, connecting the desired positions at the time t_k and t_{k+1}, by first spinning with highest possible angular velocity, and then moving along the desired trajectory (which is the line segment connecting $r_{i,HT}{}^k$ and $r_{i,HT}{}^{k+1}$) by choosing the highest possible linear velocity. However, this trivial solution only works for the ground unicycles.

Updating centroid position of the unicycle i according to Eq. (2.100) is advantageous, because transient error (difference between actual and desired centroid positions of a follower) can be reduced during evolution. However, this requires choosing $t_{k+1} - t_k$ large enough in order to ensure that the constraints on the control inputs are not violated.

Example 2.4. In this example, motion of an MAS consisting 8 follower unicycle robots and three leaders are simulated. The initial positions and orientations of the followers are listed in Tables 2.1 and 2.2. In Fig. 2.14, the paths chosen by the leaders are shown. As it is seen, leaders guide collective motion of the follower unicycles inside the narrow channel. It is noticed that desired velocities of the leaders are piecewise constant for any time period $t \in [t_k, t_{k+1}]$, where $k \in \{0, 1,\ldots, 9\}$. In Figs. 2.15 and 2.16, elements of the Jacobian Q and rigid body displacement vector D, that are obtained based on the X and Y components of positions of the leaders by applying Eq. (2.21), are depicted.

Evolution of the follower unicycles: Desired centroid positions and velocities of the followers are determined by Eq. (2.97), and each follower i updates its current centroid position by using Eq. (2.100) with $\gamma_i = \zeta_i = \eta_i = 30$. In Fig. 2.17 desired and actual positions of the follower 8 inside the narrow channel are

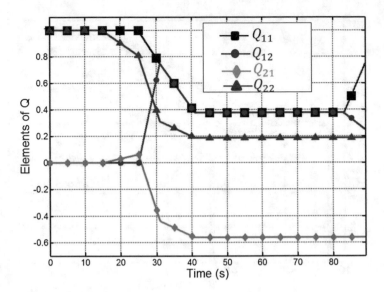

Fig. 2.15 Entries of Q versus time in Example 2.4

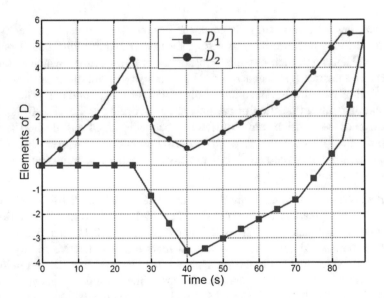

Fig. 2.16 Entries of D versus time in Example 2.4

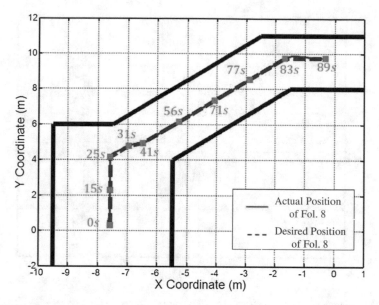

Fig. 2.17 Desired and actual paths of the centroid of follower unicycle 8

illustrated. As seen, followers reach desired way points, issued by the homogeneous transformation, at the times $t_k \in \{0, 15, 25, 31, 41, 56, 71, 77, 83, 89\}$.

As mentioned in Section 2.1.1, if $p_{i,k}(t)$ remains unchanged at any time t during MAS evolution, then actual position of each follower i is coincided on the desired position prescribed by a homogeneous mapping. In other words, $p_{i,k}(t) - \alpha_{i,k}$ quantifies deviation of a follower i from the desired state prescribed by a homogeneous mapping (the parameter $\alpha_{i,k}$, obtained from Eq. (2.18), only depends on the initial positions of the leaders and the initial centroid position of a follower i). In Fig. 2.18, parameters $p_{8,1}(t)$, $p_{8,2}(t)$, and $p_{8,3}(t)$ are depicted versus time. As illustrated, the magnitude of the parameters $p_{8,1}(t)$, $p_{8,2}(t)$, and $p_{8,3}(t)$ at the times t_k ($k = 1, 2, \ldots, 9$) are the same as the parameters, $\alpha_{8,1}$, $\alpha_{8,2}$, and $\alpha_{8,3}$, respectively. This implies that MAS configuration at t_k ($k = 1, 2, \ldots, 9$) is a homogeneous deformation of the initial configuration.

In Figs. 2.19 and 2.20, linear and angular velocities of follower unicycle robots are shown. As it is seen, u_i and ω_i do not violate the input constraint in Eq. (2.96) during evolution of the follower i.

Moreover, orientations of the follower unicycles are depicted versus time in Fig. 2.21.

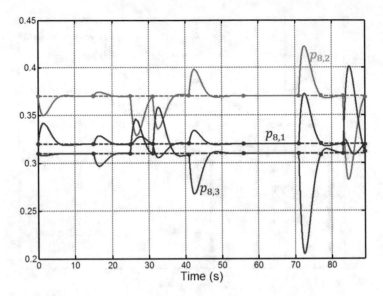

Fig. 2.18 Parameters $p_{8,1}(t)$, $p_{8,2}(t)$, and $p_{8,3}(t)$; Parameters $\alpha_{8,1}$, $\alpha_{8,2}$, and $\alpha_{8,3}$

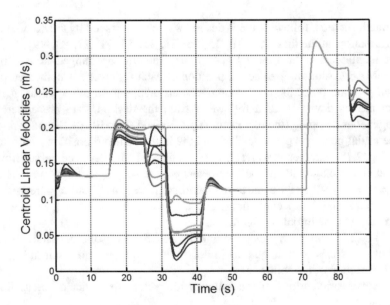

Fig. 2.19 Linear velocities of follower unicycles

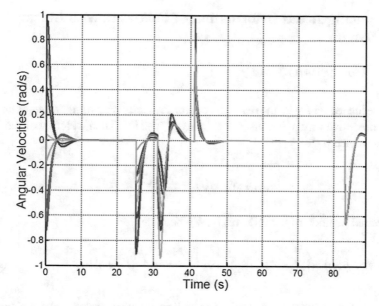

Fig. 2.20 Angular velocities of follower unicycles

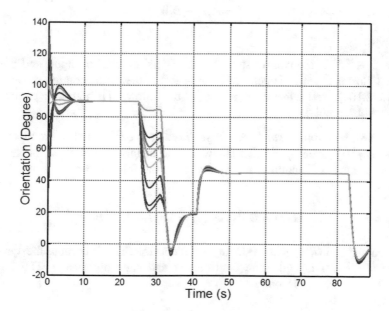

Fig. 2.21 Heading angles of the follower unicycles versus time

2.3   Homogeneous Deformation of Followers with Linear Dynamics

2.3.1   Continuous Time

Suppose the leader $j \in V_L$ moves with constant velocity on a straight line-connecting two consecutive way points at $r_j^k \in \mathbb{R}^n$ and $r_j^{k+1} \in \mathbb{R}^n$. Then, velocity of the leader j is determined by

$$\forall t \in [t_k, t_{k+1}], \ v_j(t) = \frac{r_j^{k+1} - r_j^k}{t_{k+1} - t_k} \in \mathbb{R}^n. \tag{2.102}$$

and desired position of the follower $i \in V_F$ becomes

$$\forall t \in [t_k, t_{k+1}], \ r_{i,HT}{}^k(t) = \sum_{j=1}^{n+1} \alpha_{i,j} \left[\frac{r_j^{k+1} - r_j^k}{t_{k+1} - t_k}(t - t_k) + r_j^k \right]. \tag{2.103}$$

It is also assumed that a follower i has a linear dynamics that is defined by

$$\begin{cases} \dot{Z}_i = A_i Z_i + B_i W_i \\ r_i = C_i Z_i \end{cases}, \tag{2.104}$$

where $Z_i \in \mathbb{R}^{n_z}$ is the control state, $W_i \in \mathbb{R}^{n_w}$ is the control input, and position of the follower i (which is denoted by $r_i \in \mathbb{R}^n$) is the control output of the linear system (2.104). Notice that the dynamics of the follower i is both state controllable and state observable.

Robust Tracking and Disturbance Rejection: It is desired that r_i asymptotically tracks $r_{i,HT}{}^k(t)$ given by Eq. (2.103), while the disturbance $d_i(t)$ is rejected [17]. The plant transfer function of the dynamics of the follower $i \in V_F$ is obtained as follows:

$$G_i(s) = D_i^{-1}(s)N_i(s) = C_i(sI - A_i)^{-1}B_i. \tag{2.105}$$

It is noted that $G_i(s) = D_i^{-1}(s)N_i(s)$ is left coprime. This is because the dynamics of the follower i is both state controllable and state observable. The Laplace transform of $r_{i,HT}{}^k(t)$ becomes

$$\hat{r}_{i,HT}^k(s) = 1/s^2(\mu_i^k s + v_i^k), \tag{2.106}$$

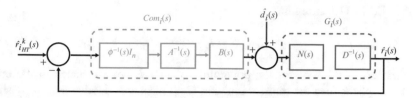

Fig. 2.22 Block diagram for robust tracking and disturbance rejection

where

$$\mu_i^k = \sum_{j=1}^{n+1} \alpha_{i,j} \left(r_j^k - t_k \frac{r_j^{k+1} - r_j^k}{t_{k+1} - t_k} \right) \tag{2.107}$$

$$v_i^k = \sum_{j=1}^{n+1} \alpha_{i,j} \frac{r_j^{k+1} - r_j^k}{t_{k+1} - t_k}. \tag{2.108}$$

Additionally, the Laplace transform of $d_i(t)$ is expressed by

$$\hat{d}_i(s) = N_{d_i} D_{d_i}^{-1}. \tag{2.109}$$

It is assumed that $\Phi_i(s) = s^2 p_i(s)$ is the least common denominator of the unstable poles $\hat{r}_{i,HT}^k(s)$ and $\hat{d}_i(s)$. Therefore, roots of $p_i(s)$ encompass unstable poles of $\hat{d}_i(s)$.

Block diagram for asymptotic tracking of desired position and disturbance rejection is shown in Fig. 2.22.

Design of the Compensator: For a given plant transfer function $G_i(s) = D_i^{-1}(s)N_i(s)$, the compensator $Com_i(s) = B(s)(\Phi(s)A(s))^{-1}$ is designed such that

1. the transfer function from $\hat{r}_{i,HT}^k(s)$ to $\hat{r}_i^k(s)$,

$$\begin{aligned} G_{0_i}(s) &= G_i(s)Com_i(s)[I_n + G_i(s)Com_i(s)]^{-1} \\ &= [I_n + G_i(s)Com_i(s)]^{-1} G_i(s)Com_i(s) \\ &= [I_n + D^{-1}NB(\Phi A)^{-1}]^{-1} D^{-1}NB(\Phi A)^{-1}, \end{aligned} \tag{2.110}$$

is asymptotically stable, and
2. the desired position $r_{i,HT}{}^k(t)$ is asymptotically tracked by $r_i(t)$ and the disturbance $d_i(t)$ is rejected.

Suppose

$$D\Phi A + NB = F \tag{2.111}$$

then Eq. (2.110) simplifies to

$$G_{0_i}(s) = I_n - \Phi A F^{-1} D. \tag{2.112}$$

Remark 2.1. If the roots of the polynomial matrix F are all placed in the open left half s-plane, then the transfer function $G_{0_i}(s)$ is stable. Notice that the desired position $r_{i,HT}{}^k(t)$ is asymptotically tracked and the disturbance $d_i(t)$ is rejected, if $\Phi_i(s) = s^2 p_i(s)$.

Example 2.5. Let each follower be a double integrator that moves in a plane, then the dynamics of the follower $i \in V_F$ becomes

$$\begin{cases} \dot{Z}_i = A_i Z_i + B_i W_i = \begin{bmatrix} 0 & 1 & 0 & 0 \\ 0 & 0 & 0 & 0 \\ 0 & 0 & 0 & 1 \\ 0 & 0 & 0 & 0 \end{bmatrix} Z_i + \begin{bmatrix} 0 & 0 \\ 1 & 0 \\ 0 & 0 \\ 0 & 1 \end{bmatrix} W_i \\ \\ r_i = C_i Z_i = \begin{bmatrix} 1 & 0 & 0 & 0 \\ 0 & 0 & 1 & 0 \end{bmatrix} Z_i \end{cases} \tag{2.113}$$

It is desired that each follower asymptotically tracks the desired position specified by Eq. (2.103).

The plant transfer function

$$G_i(s) = C_i(sI - A_i)^{-1} B_i = \begin{bmatrix} \dfrac{1}{s^2} & 0 \\ 0 & \dfrac{1}{s^2} \end{bmatrix} = D^{-1} N \tag{2.114}$$

is strictly proper. The fraction $G_i(s) = D^{-1}N$ is left coprime, where $D = s^2 I_2$ and $N = I_2$. Let

$$F(s) = D\Phi A + NB = \begin{bmatrix} (s+3)^4 & 0 \\ 0 & (s+3)^4 \end{bmatrix} \tag{2.115}$$

then $\Phi = I_2$,

$$A(s) = \begin{bmatrix} s^2 + 12s + 54 & 0 \\ 0 & s^2 + 12s + 54 \end{bmatrix},$$

and

$$B(s) = \begin{bmatrix} 108s + 1 & 0 \\ 0 & 108s + 1 \end{bmatrix}.$$

Therefore,

$$G_0(s) = \begin{bmatrix} \dfrac{108s+81}{(s+3)^4} & 0 \\ 0 & \dfrac{108s+81}{(s+3)^4} \end{bmatrix} \tag{2.116}$$

and

$$\hat{r}_{i,HT}^k(s) - \hat{r}_i^k(s) = \begin{bmatrix} \dfrac{s^4+12s^3+54s^2}{(s+3)^4} & 0 \\ 0 & \dfrac{s^4+12s^3+54s^2}{(s+3)^4} \end{bmatrix} \hat{r}_{i,HT}^k(s). \tag{2.117}$$

Consequently, the transient error $(r_{i,HT}^k(t) - r_i(t))$ converges to zero, and the desired position $r_{i,HT}^k(t)$ is asymptotically tracked by the follower $i \in V_F$ during the time interval $t \in [t_k, t_{k+1}]$.

2.3.2 Discrete Time Finite Time Reachability Model

In this subsection, it is demonstrated how followers can reach desired way points, specified by a homogeneous deformation, in a finite horizon of time.

Suppose that dynamics of the follower $i \in V_F$ is updated by

$$\begin{cases} Z_i[K+1] = A_i Z_i[K] + B_i W_i[K] + d_i[K] \\ r_i[K] = C_i Z_i[K] + n_i[K] \end{cases} \tag{2.118}$$

where $K = 1, 2, \ldots$, denotes time steps, $Z_i \in \mathbb{R}^{n_z}$, $W_i \in \mathbb{R}^{n_w}$, and $r_i \in \mathbb{R}^n$ $(n = 1, 2, 3)$ are control state, input and output, respectively. Furthermore, $d_i \in \mathbb{R}^{n_z}$ and $n_i \in \mathbb{R}^n$ are zero-mean disturbance and measurement noise, respectively. The dynamics (2.118) is controllable, therefore, the matrix

$$S_i = \begin{bmatrix} A_i^{p-1} & \ldots & B_i \end{bmatrix} \tag{2.119}$$

has the rank n_z, if $p \geq n_z$. Let $K = (J-1)p + l$ $(J = 1, 2, \ldots$ and $l = 1, 2, \ldots, p)$. Then the dynamics (2.119) can be written in the following p-step ahead form:

$$\begin{cases} Z_i[Jp] = A_i^p Z_i[(J-1)p] + S_i CO_i[(J-1)p] + D_i[(J-1)p] \\ r_i[Jp] = C_i Z_i[Jp] + n_i[Jp] \end{cases} . \tag{2.120}$$

Note that

$$D_i = \sum_{l=0}^{p-1} A_i^{p-1-l} d_i[(J-1)p+l] \in \mathbb{R}^{n_z} \tag{2.121}$$

$$CO_i = \begin{bmatrix} W_i[(J-1)p] \\ \vdots \\ W_i[Jp-1] \end{bmatrix} \in \mathbb{R}^{n_w p}. \tag{2.122}$$

Deterministic Dynamics: If disturbance and measurement noise are both zero, then it can be assured that the desired output

$$r_{i,HT}[Jp] = \sum_{k=1}^{n+1} \alpha_{i,k} r_k[Jp] \tag{2.123}$$

can be reached at $K = p,\ 2p, 3p, \ldots$. This implies that there exists a control vector CO_i^* assuring the reachability condition,

$$r_i[Jp] - r_{i,HT}[Jp] = 0. \tag{2.124}$$

at $J = 1,\ 2, \ldots$. The control CO_i^* also minimizes the cost function

$$cost = \frac{1}{2} CO_i[(J-1)p]^T \Omega_{J-1} CO_i[(J-1)p]$$

$$+ \sum_{l=0}^{p-1} \left[r_i[(J-1)p+l] - r_{i,HT}[(J-1)p+l] \right]^T \Gamma_{J-1} \left[r_i[(J-1)p+l] - r_{i,HT}[(J-1)p+l] \right]. \tag{2.125}$$

where $\Omega_{J-1} \in \mathbb{R}^{n_w \times n_w}$ and $\Gamma_{J-1} \in \mathbb{R}^{n \times n}$ are positive definite weight matrices.

Equation (2.125) can be rewritten as

$$cost = \frac{1}{2} CO_i[(J-1)p]^T (\Omega_{J-1} + P_{J-1}) CO_i[(J-1)p] + E_{J-1}^T CO_i[(J-1)p]$$

$$+ \sum_{l=1}^{p} (C_i A_i^l Z[(J-1)p]$$

$$- r_{i,HT}[(J-1)p+l]) \Gamma_{J-1} (C_i A_i^l Z[(J-1)p] - r_{i,HT}[(J-1)p+l]). \tag{2.126}$$

where the jth $(j = 1, 2, \ldots, p)$ block of $E_{J-1} \in \mathbb{R}^{p.n_w}$ and the $i_1 i_2$ $(i_1, i_2 = 1, 2, \ldots, p)$ block of $P_{J-1} \in \mathbb{R}^{p.n_w \times n.n_w}$ are obtained as follows:

$$E_{J-1} = 2 \sum_{\substack{l=1 \\ j \leq l}}^{p} \left(C_i A_i^{l-j} B_i\right)^T \Gamma_{J-1} \left(C_i A_i^l B_i Z_i[(J-1)p] - r_{i,HT}[(J-1)p+l]\right) \in \mathbb{R}^{n_w \times n_w}$$

$$(2.127)$$

$$\{P_{J-1}\}_{i_1 i_2} = \sum_{l=max\{i_1,i_2\}}^{p} \left(C_i A_i^{l-i_1} B_i\right)^T \Gamma_{J-1} \left(C_i A_i^{l-i_2} B_i\right) \in \mathbb{R}^{n_u \times n_u}. \qquad (2.128)$$

To specify CO_i^* the augmented cost function

$$COST = \frac{1}{2} CO_i[(J-1)p]^T (\Omega_{J-1} + P_{J-1}) CO_i[(J-1)p] +$$

$$E_J^T CO_i[(J-1)p] + \sum_{l=1}^{p} (C_i A_i^l Z[(J-1)p] -$$

$$r_{i,HT}[(J-1)p+l])^T \Gamma_{J-1} (C_i A_i^l Z_i[(J-1)p] - r_{i,HT}[(J-1)p+l]) +$$

$$\lambda_i[J-1]^T \left(r_{i,HT}[Jp] - C_i(A_i^p Z_i[(J-1)p] + S_i CO_i[(J-1)p])\right))$$

$$(2.129)$$

should be minimized, where $\lambda_i[J-1] \in \mathbb{R}^n$ is the Lagrange multiplier vector. Therefore,

$$CO_i = CO_i^* = (\Omega_{J-1} + P_{J-1})^{-1}(S_i^T C_i \lambda_i[J-1]^* - E_{J-1}) \qquad (2.130)$$

with

$$\lambda_i[J-1] = \lambda_i[J-1]^* = (C_i S_i (\Omega_{J-1} + P_{J-1})^{-1} S_i^T C_i^T)^{-1}$$

$$\times (r_{i,HT}[Jp] + C_i S_i (\Omega_{J-1} + P_{J-1})^{-1} E[J-1] - C_i A_i^p Z[(J-1)p]) \qquad (2.131)$$

minimizes the cost (2.126), and finite time reachability of the desired position $r_{i,HT}$ is guaranteed.

Stochastic Dynamics: It is assumed D_i and n_i represent nonzero zero-mean disturbance and measurement noise, respectively, where they are uncorrelated and

$$\mathbb{E}(D_i^T D_i) = Q_i \qquad (2.132)$$

$$\mathbb{E}(n_i^T n_i) = R_i. \qquad (2.133)$$

denote the expected values of $D_i^T D_i$ and $n_i^T n_i$, respectively. In presence of disturbance and measurement noise, Kalman filter [130] is used to estimate the state Z_i. Let $\hat{Z}_i^-$ and $\hat{Z}_i^+$ denote prediction and measurement update of the state Z_i. Then,

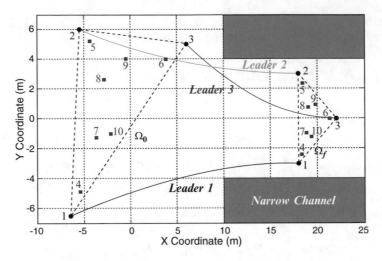

Fig. 2.23 Leaders' paths in Example 2.6

$$\hat{CO}_i = (\Omega_{J-1} + P_{J-1})^{-1}(S_i^T C_i \hat{\lambda}_i[J-1]^* - E_{J-1})$$

$$\hat{\lambda}_i[J-1] = \lambda_i[J-1]^* = (C_i S_i (\Omega_{J-1} + P_{J-1})^{-1} S_i^T C_i^T)^{-1}$$

$$\times (r_{i,HT}[Jp] + C_i S_i (\Omega_{J-1} + P_{J-1})^{-1} E[J-1] - C_i A_i^p \hat{Z}_i^+[(J-1)p])$$

(2.134)

and states are estimated by using Kalman filter as follows:

$$\begin{cases} P_i^-[Jp] = A_i^p P_i^+[(J-1)p] A_i^{pT} \\ K_i[Jp] = P_i^-[Jp] C_i^T (C_i P_i^-[Jp] C_i^T + R_i) \\ \hat{Z}_i^-[Jp] = A_i^p \hat{Z}_i^+[(J-1)p] + S_i \hat{CO}_i \\ \hat{Z}_i^+[Jp] = \hat{Z}_i^-[Jp] + K_i[Jp](r_{i,HT}[Jp] - C_i \hat{Z}_i^-[Jp]) \\ P_i^+[Jp] = (I - K_i[Jp]C_i)P_i^-(I - K_i[Jp]C_i)^T + K_i[Jp]R_i K_i^T[Jp] \end{cases}$$

(2.135)

Example 2.6 ([105]). Consider an MAS consisting of 10 agents (3 leaders and 7 followers) with the initial and final formations shown in Fig. 2.23. Leaders choose the paths shown in Fig. 2.23, where they start their motion from rest at $K = 1$ and they stop at $K = 3000$.

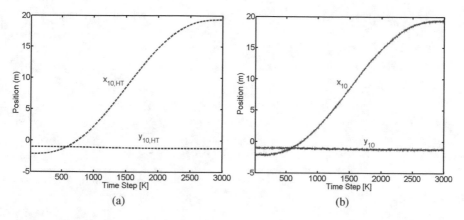

Fig. 2.24 (a) X and Y components of the desired position $r_{10,HT} = x_{10,HT}\hat{\mathbf{e}}_x + y_{10,HT}\hat{\mathbf{e}}_y$; (b) actual positions $r_{10} = x_{10}\hat{\mathbf{e}}_x + y_{10}\hat{\mathbf{e}}_y$

It is assumed that followers have the linear dynamics defined by Eq. (2.120), where

$$
A_i = \begin{bmatrix} 1 & 1 & 0 & 0 \\ 0 & 1 & 0 & 0 \\ 0 & 0 & 1 & 1 \\ 0 & 0 & 0 & 1 \end{bmatrix}
$$

$$
B_i = \begin{bmatrix} 0 & 0 \\ 1 & 0 \\ 0 & 0 \\ 0 & 1 \end{bmatrix}
$$

$$
C_i = \begin{bmatrix} 1 & 0 & 0 & 0 \\ 0 & 0 & 1 & 0 \end{bmatrix}.
$$

and $K = (J-1)p + l = 1,\ 2,\ \ldots,\ 3000$ ($p = 4$, $J = 1,\ 2,\ldots,\ 750$). Followers update their states according to Eqs. (2.134) and (2.135), where disturbance and measurement noise are characterized by $Q_i = 0.1I \in \mathbb{R}^{4\times4}$ and $R_i = 0.1I \in \mathbb{R}^{2\times2}$.

Given leaders' positions at $K = 1,\ 2,\ldots,\ 3000$, desired position of the follower 10 is obtained from Eq. (2.17). Shown in Fig. 2.24(a) are the X and Y components of the desired position $r_{10,HT} = x_{10,HT}\hat{\mathbf{e}}_x + y_{10,HT}\hat{\mathbf{e}}_y$. Furthermore, components of actual position $r_i = x_{10}\hat{\mathbf{e}}_x + y_{10}\hat{\mathbf{e}}_y$ are illustrated versus time in Fig. 2.24(b).

Chapter 3
Homogeneous Deformation of Multi-Agent Systems Communication

In this chapter, it is shown how a multi-agent system (MAS) can acquire a desired homogeneous deformation in $\mathbb{R}^n$ (prescribed by $n+1$ leaders) through local communication. For this purpose, two communication protocols are developed. The first protocol, that is called *minimum communication*, allows each follower to communicate only with $n+1$ local agents. Under this protocol, communication weights are uniquely determined based on the initial positions of a follower i and $n+1$ agents that are adjacent to the follower i. Followers apply first order linear dynamics to acquire the desired position defined by a homogeneous transformation. The second protocol, that is called *preservation of volumetric ratios*, permits followers to interact with more than $n+1$ local agents. Under this setup, local volumetric ratios are obtained based on the initial positions of a follower i and $m_i \geq n+1$ agents that are adjacent to the follower i. Then each follower applies a nonlinear dynamics to acquire the desired position (prescribed by a homogeneous mapping) through keeping transient volumetric ratios, specified based on current positions of the agents, as close as initial values of the volumetric ratios.

3.1 Graph Theory Notions and Definitions

Basic Notions: For an MAS moving in $\mathbb{R}^n$ (n denotes dimension of the motion field; n can be either 1, 2, or 3.), interagent communication is prescribed by a directed graph (digraph) $G = (V_G, E_G)$, where $V_G = \{1, 2, \ldots, N\}$ and $E_G \in V_G \times V_G$ are sets of nodes and edges of the graph. The state (position and velocity) of the node i can be accessed by the node j, if $(i,j) \in E_G$. The set $N_i = \{j : (j,i) \in E_G\}$ is called the *in-neighbor set of the vertex* i, where $d_i = |N_i|$ is the cardinality of N_i. A graph is called undirected, if communication between every two connected nodes of the graph is bidirectional. (If $(j,i) \in E_G$, then $(i,j) \in E_G$.) A finite (or infinite) sequence of edges which connects a finite sequence of vertices is called a

© Springer International Publishing AG 2016
H. Rastgoftar, *Continuum Deformation of Multi-Agent Systems*,
DOI 10.1007/978-3-319-41594-9_3

path. An undirected graph is connected, if there exists at least a path between any two nodes of an undirected graph. A directed graph is called *weakly connected*, if substituting every directed edge by an undirected edge yields a connected undirected graph. Two vertices i and j in the digraph G are called *connected*, if there exists a directed path from i to j, and one from j to i. A digraph G is called *strongly connected*, if every two nodes of G is connected. A graph G defining interagent communication can be expressed by the independent sum

$$G = \phi \oplus \partial\phi,$$

where the $\partial\phi$ is the boundary graph representing boundary nodes of G and the subgraph ϕ represents interior nodes of G.

Definition 1. An agent i is called a *leader*, if the in-neighbor set N_i is empty. This implies that leaders move independently. Leaders are identified by the numbers $1, 2, \ldots, N_l$ ($N_l \leq N$). The leader set

$$V_L = \{1, 2, \ldots, N_l\}$$

defines index numbers of leaders.

Definition 2. An agent i is called a follower, if the in-neighbor set N_i is nonempty. In other words, every follower agent can access state information (position and velocity) of the in-neighbor agents. Followers are identified by the numbers $N_l + 1, N_l + 2, \ldots, N$. The follower set

$$V_F = \{N_l + 1, N_l + 2, \ldots, N\}$$

defines index numbers of followers.

Definition 3. The set $C \subset \mathbb{R}^n$ is convex, if for every x and y in C, then $(1 - \alpha)x + \alpha y$ is in C for all $\alpha \in [0, 1]$.

Definition 4. Let $R = \{r_1, r_2, \ldots, r_{N_l}\}$ specify a set of vectors in $\in \mathbb{R}^n$, then intersections of convex sets containing R is called a convex hull. Mathematically speaking,

$$Conv(R) = \{r_i = \sum_{j=1}^{N_l} \alpha_{i,j}, r_j \in \mathbb{R}^n | \alpha_{i,j} \geq 0 \wedge \sum_{j=1}^{N} \alpha_{i,j} = 1\}. \qquad (3.1)$$

3.2   Protocol of Minimum Communication

Consider a communication graph $G = \phi \oplus \partial\phi$ with $n + 1$ nodes belonging to the boundary graph $\partial\phi$ and $N - n - 1$ nodes belonging to the subgraph ϕ. The nodes belonging to the subgraph ϕ and the boundary graph $\partial\phi$ represent followers

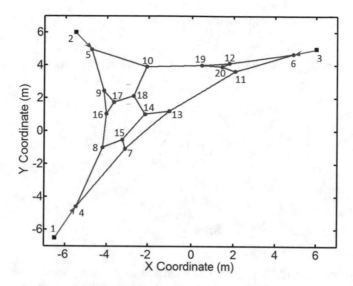

Fig. 3.1 An interagent communication graph used for MAS evolution in a plane ($\in \mathbb{R}^2$)

and leaders, respectively. Notice that leaders move independently, but position of each leader is tracked by a follower. Therefore, follower-leader communication is unidirectional and shown by an arrow terminated to the follower. Communication between two follower agents can be either unidirectional or bidirectional. This implies that the subgraph ϕ is directed. It is noted that communication weights are not inevitably the same for the two connected nodes in ϕ. Also, any node belonging to the subgraph ϕ can access state information of $n+1$ in-neighbor agents. A typical communication graph for an MAS evolving in a plane ($\in \mathbb{R}^2$) is shown in Fig. 3.1. As it is seen in Fig. 3.1, the MAS consists of 20 agents (3 leaders and 17 followers), where each follower interacts with 3 local agents, and communication between two in-neighbor followers is bidirectional.

Next, it is described how communication weights can be uniquely determined based on the initial positions of the agents.

3.2.1 Communication Weights

Let the follower $i \in V_F$ access positions of $n+1$ agents i_1, i_2,..., i_{n+1} belonging to the in-neighbor set N_i. Suppose that agents i, i_1, i_2, ..., i_{n+1} are initially positioned at R_i, R_{i_1}, R_{i_2}, ..., $R_{i_{n+1}}$, respectively. It is assumed that

$$Rank \left[R_{i_2} - R_{i_1} \;\ldots\; R_{i_{n+1}} - R_{i_1} \right] = n. \tag{3.2}$$

Then R_i can be uniquely expanded as follows:

$$R_i = R_{i_1} + \sum_{j=2}^{n+1} w_{i,i_j}(R_{i_j} - R_{i_1}) = (1 - \sum_{j=2}^{n+1} w_{i,i_j})R_{i_1} + \sum_{j=2}^{n+1} w_{i,i_j}R_{i_j}. \tag{3.3}$$

Let

$$w_{i,i_1} = 1 - \sum_{j=2}^{n+1} w_{i,i_j}, \tag{3.4}$$

then

$$R_i = \sum_{j=1}^{n+1} w_{i,i_j}R_{i_j}. \tag{3.5}$$

By considering Eq. (3.4), and Eq. (3.5) written in the component wise form, the parameter w_{i,i_j} is uniquely determined by solving the following set of $n+1$ linear algebraic equations:

$$\begin{bmatrix} X_{1,i_1} & X_{1,i_2} & \cdots & X_{1,i_{n+1}} \\ X_{2,i_1} & X_{2,i_2} & \cdots & X_{2,i_{n+1}} \\ \vdots & \vdots & & \vdots \\ X_{n,i_1} & X_{n,i_2} & \cdots & X_{n,i_{n+1}} \\ 1 & 1 & & 1 \end{bmatrix} \begin{bmatrix} w_{i,i_1} \\ w_{i,i_2} \\ \vdots \\ w_{i,i_n} \\ w_{i,i_{n+1}} \end{bmatrix} = \begin{bmatrix} X_{1,i} \\ X_{2,i} \\ \vdots \\ X_{n,i} \\ 1 \end{bmatrix}. \tag{3.6}$$

It is noted that $X_{q,h}$ is the qth component of the initial position of the agent $h \in \{i, i_1, \ldots, i_{n+1}\}$. The communication weights are all positive, if the follower $i \in V_F$ is initially located inside the communication polytope whose vertices are occupied by the agents $i_1, \ldots, i_{n+1}$.

MAS evolution in a plane: If an MAS evolves in a plane, then homogeneous deformation is prescribed by three leaders at the vertices of the leading triangle; desired position of a follower i is specified by Eq. (2.17) as a convex combination of positions of three leaders. Follower i interacts with three agents i_1, i_2, and i_3 to acquire the desired position through local communication. As shown in Fig. 3.2, the motion plane is divided into seven subregions based on the signs of the communication weights. It is evident that communication weights w_{i,i_1}, w_{i,i_2}, and w_{i,i_3} are all positive, if the follower i is initially placed inside the communication triangle whose vertices are occupied by the adjacent agents i_1, i_2, and i_3.

For MAS evolution in the $X - Y$ plane, communication weights simplify to

$$\begin{bmatrix} X_{i_1} & X_{i_2} & X_{i_3} \\ Y_{i_1} & Y_{i_2} & Y_{i_3} \\ 1 & 1 & 1 \end{bmatrix} \begin{bmatrix} w_{i,i_1} \\ w_{i,i_2} \\ w_{i,i_3} \end{bmatrix} = \begin{bmatrix} X_i \\ Y_i \\ 1 \end{bmatrix}. \tag{3.7}$$

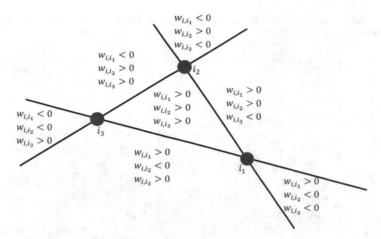

Fig. 3.2 Dividing motion plane into seven sub-regions based on the signs of communication weights

where the first and second components of the initial position of the agent $h \in \{i, i_1, i_2, i_3\}$ are denoted by X_h and Y_h, respectively. In Table 3.1, positions of agents in the initial formation, shown in Fig. 3.1, and the corresponding communication weights are listed.

Remark 3.1. It is noticed that the total number of the agents is even, when an MAS is supposed to evolve in a plane and the minimum interagent communication is issued by a connected (and undirected) graph. This is because the MAS consists of 3 leaders, and each follower is restricted to interact with three local agents, where follower-follower communication is bidirectional.

3.2.2   Weight Matrix

Let the weight matrix $W \in \mathbb{R}^{(N-n-1) \times (N-n-1)}$ be defined as follows:

$$W_{ji} = \begin{cases} w_{i+n+1,j} > 0 & if\,(j, i+n+1) \in E_G \\ -1 & i+n+1 = j \\ 0, & else \end{cases}. \qquad (3.8)$$

Then matrix W has the following properties:

- Each row of the matrix W has $n+2$ nonzero entries.
- Sum of every row of the matrix W is zero.

Table 3.1 Initial positions and communication weights associated with the formation shown in Fig. 3.1

Index i	$X_i(m)$	$Y_i(m)$	i_1	i_2	i_3	w_{i,i_1}	w_{i,i_2}	w_{i,i_3}	$X_{i,F}(m)$	$Y_{i,F}(m)$
1	−6.5000	−6.500	–	–	–	–	–	–	6.0000	11.5000
2	−5.5000	6.0000	–	–	–	–	–	–	0.0000	11.5000
3	6.0000	5.0000	–	–	–	–	–	–	6.0000	16.0000
4	−5.4776	−4.5886	1	7	8	0.65	0.20	0.15	5.4935	11.8323
5	−4.7379	4.9601	2	9	10	0.65	0.21	0.14	0.9018	11.8253
6	4.8628	4.6781	3	11	12	0.72	0.15	0.13	5.6246	15.5646
7	−3.1087	−1.0772	4	13	15	0.29	0.36	0.35	4.8046	12.6445
8	−4.2060	−0.9880	4	15	16	0.28	0.29	0.43	4.2353	12.2154
9	−4.1478	2.4466	5	16	17	0.31	0.42	0.27	2.4846	12.1321
10	−2.0849	3.9020	5	18	19	0.35	0.23	0.42	2.7144	12.8888
11	2.0673	3.6233	6	13	20	0.33	0.20	0.47	4.8381	14.5108
12	1.7904	4.1125	6	19	20	0.22	0.41	0.37	4.4527	14.3881
13	−1.0684	1.2105	7	11	14	0.32	0.33	0.35	4.5926	13.3668
14	−2.1595	1.0272	13	15	18	0.42	0.27	0.31	4.1674	12.9485
15	−3.2445	−0.5208	7	8	14	0.41	0.34	0.25	4.4517	12.5746
16	−4.0265	1.0415	8	9	17	0.35	0.36	0.29	3.2700	12.2226
17	−3.6591	1.7465	9	16	18	0.27	0.43	0.30	3.0800	12.3436
18	−2.6927	2.1270	10	14	17	0.29	0.34	0.37	3.3437	12.7074
19	0.4589	3.9924	10	12	20	0.32	0.42	0.26	3.8802	13.8745
20	1.4389	3.9094	11	12	19	$\frac{1}{3}$	$\frac{1}{3}$	$\frac{1}{3}$	4.3904	14.2578

If the matrix W is partitioned as follows:

$$W = \left[B \in \mathbb{R}^{(N-n-1)\times(n+1)} \quad A \in \mathbb{R}^{(N-n-1)\times(N-n-1)} \right], \tag{3.9}$$

then every column of the matrix B has only one positive entry, while the remaining entries of B are all zero. Even if follower-follower communication is bidirectional, the matrix A is not necessarily symmetric; however, if $A_{ij} = 0$, then $A_{ji} = 0$, and if $A_{ij} \neq 0$, then $A_{ji} \neq 0$.

In the following theorem, it is proven that the matrix $A = -(I - F)$ is Hurwitz, if the matrix $F \in \mathbb{R}^{(N-n-1)\times(N-n-1)}$ is nonnegative and irreducible.

Theorem 3.1. *For MAS evolution in $\mathbb{R}^n$, if (i) interagent communication is defined by a digraph $G = \phi \oplus \partial\phi$ with directed and strongly connected subgraph ϕ, and (ii) communication weights, determined by Eq. (3.6), are all positive, then the matrix A is Hurwitz.*

Proof. The matrix $A = -(I - F)$ is obtained by eliminating the first $n + 1$ columns from the matrix W. Since W is zero-sum row, sum of each of the first $n + 1$ rows of A is negative, and the remaining rows of A are zero-sum. In other words, sum

Fig. 3.3 An initial
distribution of agents
resulting in negative weights
of communication for some
followers

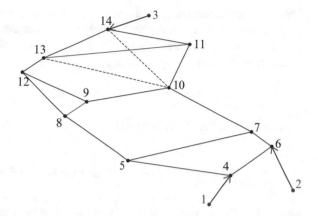

of each of the first $n+1$ rows of the nonnegative and irreducible matrix F is less
than 1, while the remaining rows of F are one-sum. By provoking Perron-Frobenius
theorem [102], it is concluded that the spectrum of the matrix F, which is denoted
by $\rho(F)$, is not greater than 1. However, A is a non-singular M-matrix [9]. This is
because sum of each of the first $n+1$ rows of the matrix A is negative. Therefore,
the spectrum of the matrix F cannot be equal to 1 and the matrix A is necessarily
Hurwitz.

Remark 3.2. It is not necessarily required that all communication weights are
positive in order to assure achieving homogeneous transformation under local
communication. In fact, positiveness of communication weights is a sufficient
condition. For instance, initial distribution of an MAS shown in Fig. 3.3 results
in two positive communication weights $w_{11,10}$ and $w_{11,14}$ and a negative weight
$w_{11,13}$. However, the partition A of the weight matrix W that is consistent with the
initial positions of agents is still Hurwitz. This guarantees that positions of followers
asymptotically converge to desired positions in the ultimate formation, where final
formation is a homogeneous transformation of the initial configuration.

***Desired Positions of Followers Issued by a Homogeneous Deformation*:** Let

$$Z_q = \left[X_{q,n+2} \ \ldots \ X_{q,N} \right]^T \in \mathbb{R}^{N-n-1}$$

and

$$U_q = \left[X_{q,1} \ \ldots \ X_{q,n+1} \right]^T \in \mathbb{R}^{n+1}$$

denote the qth components of the initial positions of the followers and leaders,
respectively, then

$$AZ_q + BU_q = 0. \tag{3.10}$$

This represents consistency between communication weights and initial positions of the agents (See Eq. (3.6).) Notice that the qth component of Eq. (3.3) is the same as the row $i - n - 1$ of Eq. (3.10). Because A is Hurwitz, the qth components of positions of followers are determined by

$$Z_q = -A^{-1} B U_q. \tag{3.11}$$

The row i of Eq. (3.11) is equal to

$$X_{q,i+n+1} = \sum_{k=1}^{n+1} \alpha_{i+n+1,k} X_{q,k}(t), \tag{3.12}$$

where $X_{q,i}$ (the qth component of the position of the follower $i \in V_F$) is expressed as the linear combination of the qth components of positions of leaders. It is noted that the parameter $\alpha_{i+n+1,k}$ is uniquely obtained from Eq. (2.15), if initial positions of leaders satisfy the rank condition (2.3). Therefore, the ik entry of the matrix

$$W_L = -A^{-1} B \in \mathbb{R}^{(N-n-1) \times (n+1)}. \tag{3.13}$$

is inevitably equal to $\alpha_{i+n+1,k}$.

The desired position of the follower i, satisfying the condition of a homogeneous transformation, is specified by Eq. (2.14). Let

$$x_{q,i+n+1,HT} = \sum_{k=1}^{n+1} \alpha_{i+n+1,k} x_{q,k}(t), \ (i+n+1) \in V_F, \tag{3.14}$$

be the qth component of $r_{i+n+1,HT}$, then

$$z_{q,HT}(t) = \left[x_{q,n+2,HT} \ \cdots \ x_{q,N,HT} \right]^T \in \mathbb{R}^{N-n-1}$$

defines the qth components of the desired positions of followers. Given $u_q(t)$ (the qth components of the leaders' positions at a time t), then

$$z_{q,HT}(t) = W_L u_q(t). \tag{3.15}$$

Remark 3.3. If communication weights are all positive, then Eq. (3.15) assures that every follower is initially placed inside the ith communication polytope whose vertices are occupied by the agents $i_1, i_2, \ldots, i_{n+1}$. However, an arbitrary initial distribution of agents may not necessarily result in positive communication weights (See Remark 3.2.) In order to guarantee that positive communication weights are consistent with agents' positions, initial positions of the followers can be determined by applying the following procedure:

- A communication graph G satisfying the properties specified in Section 3.1 is considered.
- Given the communication graph G, $n + 1$ arbitrary positive weights w_{i,i_1}, w_{i,i_2}, ..., $w_{i,i_{n+1}}$ are considered for communication between the follower i and $n + 1$ in-neighbor agents $i_1, i_2, \ldots, i_{n+1}$ such that

$$\sum_{k=1}^{n+1} w_{i,i_k} = 1, \ \forall \, i \in V_F. \tag{3.16}$$

- The weight matrix $W \in \mathbb{R}^{(N-n-1) \times N}$ is set up by using the relation (3.8), and then partitions $B \in \mathbb{R}^{(N-n-1) \times (n+1)}$ and $A \in \mathbb{R}^{(N-n-1) \times (N-n-1)}$ of W are determined.
- Leaders are positioned at $R_1, R_2, \ldots, R_{n+1}$ at the initial time t_0, where leaders' initial positions satisfy the rank condition (2.3).
- The qth components of the initial positions of followers are assigned by using Eq. (3.11).

Example 3.1. Let the graph shown in Fig. 3.4 define communication among agents of an MAS containing 3 leaders and 7 followers. The boundary nodes 1, 2, and 3 represent leaders, and the interior nodes 4, 5, ..., 10 represent followers. As it is seen, each follower interacts with three local agents, where follower-follower communication is bidirectional.

Weights listed in Table 3.2 are considered for inter-agent communications among the agents. Notice that sum of the communication weights of each follower is equal

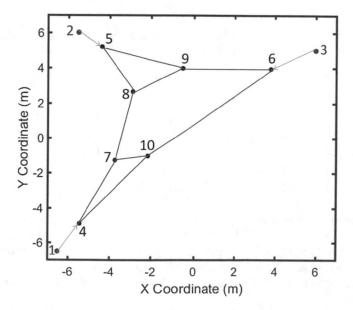

Fig. 3.4 A typical communication graph used for MAS evolution in a plane ($\in \mathbb{R}^2$)

Table 3.2 Communication
weights associated with the
formation shown in Fig. 3.4

Index i	i_1	i_2	i_3	w_{i,i_1}	w_{i,i_2}	w_{i,i_3}
1	–	–	–	–	–	–
2	–	–	–	–	–	–
3	–	–	–	–	–	–
4	1	7	10	0.70	0.15	0.15
5	2	8	9	0.70	0.15	0.15
6	3	9	10	0.70	0.15	0.15
7	4	8	10	0.40	0.36	0.24
8	5	7	9	$\frac{1}{3}$	$\frac{1}{3}$	$\frac{1}{3}$
9	5	6	8	0.31	0.42	0.27
10	4	6	7	0.35	0.29	0.36

Table 3.3 Initial positions of
agents; parameters
$\alpha_{i,1}$, $\alpha_{i,2}$, $\alpha_{i,3}$ for the
follower agent i

Index i	$X_i(m)$	$Y_i(m)$	α_{i,i_1}	α_{i,i_2}	α_{i,i_3}
1	−6.5000	−6.5000	–	–	–
2	−5.5000	6.0000	–	–	–
3	6.0000	5.0000	–	–	–
4	−5.4278	−4.8900	0.8647	0.0538	0.0815
5	−4.3598	5.1931	0.0562	0.8397	0.1040
6	3.8002	3.9443	0.0991	0.0836	0.8173
7	−3.7164	−1.2494	0.5636	0.2323	0.2041
8	−2.8688	2.6411	0.2487	0.5009	0.2504
9	−0.5300	3.9796	0.1262	0.4307	0.4431
10	−2.1356	−1.0174	0.5343	0.1267	0.3390

to 1 ($w_{i,i_1} + w_{i,i_2} + w_{i,i_3} = 1$, $\forall i \in V_F$). By using the definition (3.8), the weight matrix $W \in \mathbb{R}^{7 \times 10}$ is obtained as follows:

$$W = \begin{bmatrix} 0.70 & 0 & 0 & -1 & 0 & 0 & 0.15 & 0 & 0 & 0.15 \\ 0 & 0.70 & 0 & 0 & -1 & 0 & 0 & 0.15 & 0.15 & 0 \\ 0 & 0 & 0.70 & 0 & 0 & -1 & 0 & 0 & 0.15 & 0.15 \\ 0 & 0 & 0 & 0.40 & 0 & 0 & -1 & 0.36 & 0 & 0.24 \\ 0 & 0 & 0 & 0 & 1/3 & 0 & 1/3 & -1 & 1/3 & 0 \\ 0 & 0 & 0 & 0 & 0.31 & 0.42 & 0 & 0.27 & -1 & 0 \\ 0 & 0 & 0 & 0.35 & 0 & 0.29 & 0.36 & 0 & 0 & -1 \end{bmatrix}. \tag{3.17}$$

Leader agents are placed at $(X_1, Y_1) = (-6.5, -6.5)$, $(X_2, Y_2) = (-5.5, 6)$, and $(X_3, Y_3) = (6, 5)$ at the initial time. Note that initial positions of leaders satisfy the rank condition (2.3). Initial positions of the followers are obtained by using Eq. (3.11) as listed in Table 3.3. Additionally, parameters $\alpha_{i,k}$ ($\forall i \in V_F$, $k \in V_L$) are calculated by using Eq. (2.19) and listed in the last three columns of Table 3.3.

3.2.3   MAS Evolution Dynamics-First Order Kinematic Model-Method 1

Let position of the agent $i \in V$ be updated by

$$\dot{r}_i(t) = u_i \tag{3.18}$$

where

$$u_i = \begin{cases} Given & i \in V_L \\ -g_i r_i(t - h_{1i}) + g_i r_{i,d}(t - h_{2i}) + \beta \dot{r}_{i,d}(t - h_{2i}) & i \in V_F \end{cases}. \tag{3.19}$$

It is noted that leaders move independently, therefore their positions are known at any time t. Also,

$$r_{i,d}(t) = \sum_{j \in N_i} w_{i,j} r_j(t) \tag{3.20}$$

where $w_{i,j}$ is the communication weight, $g_i \in \mathbb{R}_+$ is constant, and $h_{1i} \geq 0$ and $h_{2i} \geq 0$ are constant time delays. If $h_{1i} = 0$, the follower i can immediately access its own position (without time delay) at any time t. If $h_{2i} = 0$, follower i accesses positions of its in-neighbor agents (belonging to the set N_i) without communication delay. The parameter β can be either 0 or 1. If $\beta = 0$, then position of the follower i is updated based on the positions of the neighboring agents. If $\beta = 1$, then position of the follower i is updated based on both positions and velocities of the neighboring agents. Evolution of an MAS in absence/presence of communication delays is studied in the subsections 3.2.3.1 and 3.2.3.2.

3.2.3.1   Evolution of Followers in Absence of Communication Delays

Followers without Body Size

Let h_{1i} and h_{2i} in Eqs. (3.18) and (3.19) be both zero, then position of the follower $i \in V_F$ is updated by

$$\dot{r}_i = g_i(r_{i,d}(t) - r_i(t)). \tag{3.21}$$

Suppose that $z_q = [x_{q,n+2} \ \ldots \ x_{q,N}]^T \in \mathbb{R}^{N-n-1}$ defines the qth components of the positions of followers at the time t, then z_q is updated by the following first order matrix dynamics:

$$\dot{z}_q = G(Az_q + Bu_q). \tag{3.22}$$

It is noticed that Eq. (3.21) and the row $i - n - 1$ of the MAS evolution dynamics given by Eq. (3.22) are identical, and

$$G = diag(g_{n+2}, g_{n+3}, \ldots, g_N) \in \mathbb{R}^{(N-n-1)\times(N-n-1)} \tag{3.23}$$

is a positive diagonal matrix. Because the matrix A is Hurwitz, the dynamics of Eq. (3.22) is stable and z_q asymptotically converges to

$$Z_{F,q} = W_L U_{F,q} = -A^{-1} B U_{F,q}, \tag{3.24}$$

where $U_{F,q}$ specifies the qth components of final positions of leaders. Because the ik entry of the matrix W_L is equal to $\alpha_{i+n+1,k}$ ($\alpha_{i+n+1,k}$ is uniquely determined by Eq. (2.15) based on the initial positions of the follower $i + n + 1$ and $n + 1$ leaders), final position of the follower i satisfies Eq. (3.14). Therefore, final formation of the MAS is a homogeneous deformation of the initial configuration.

Example 3.2. It is assumed that an MAS, consisting of 20 agents (3 leaders and 17 followers), negotiates a narrow channel in the $X - Y$ plane. Initial and final formations P and Q are shown in Fig. 3.5. Notice that the formation Q is a homogeneous transformation of the initial MAS configuration P, where initial positions of the agents are listed in Table 3.1. Also, final positions of the agents are listed in the last two columns of Table 3.1.

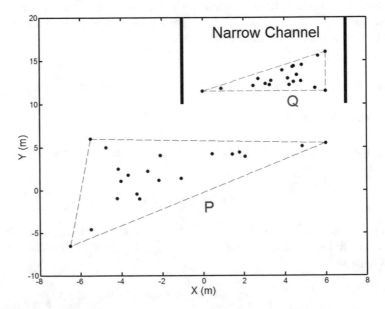

Fig. 3.5 Initial and desired final configurations of an MAS negotiating a narrow channel

Positions of the leaders are defined by

$$
\text{Leader 1}: \begin{cases} x_1(t) = -6.5 + \dfrac{12.5}{30}t & t \le 30 \\ x_1(t) = 6 & t > 30 \\ y_1(t) = -6.5 + \dfrac{18}{30}t & t \le 30 \\ y_1(t) = 11.5 & t > 30 \end{cases} \tag{3.25}
$$

$$
\text{Leader 2}: \begin{cases} x_2(t) = -5.5 + 5.5\cos(\dfrac{\pi t}{60} + \dfrac{3\pi}{2}) & t \le 30 \\ x_2(t) = 0 & t > 30 \\ y_2(t) = 11.5 + 5.5\sin(\dfrac{\pi t}{60} + \dfrac{3\pi}{2}) & t \le 30 \\ y_2(t) = 11.5 & t > 30 \end{cases} \tag{3.26}
$$

$$
\text{Leader 3}: \begin{cases} x_3(t) = 6\dfrac{12.5}{30}t & t \le 30 \\ x_3(t) = 6 & t > 30 \\ y_3(t) = 5.5 + \dfrac{10.5}{30}t & t \le 30 \\ y_3(t) = 16 & t > 30 \end{cases} \tag{3.27}
$$

In Fig. 3.6, paths chosen by the three leaders (at the vertices of the leading triangle) are depicted.

Homogeneous transformation of the MAS is related to the trajectories chosen by the three leaders placed at the vertices of the leading triangle. Entries of the Jacobian matrix $Q \in \mathbb{R}^{2\times2}$ and rigid body displacement vector $D \in \mathbb{R}^2$, that are specified based on the first (X) and second (Y) components of the positions of leaders, are shown versus time in Fig. 3.7. As it is observed, $Q(0) = I \in \mathbb{R}^{2\times2}$ and $D(0) = 0 \in \mathbb{R}^2$. Therefore, $r_i(0) = R_i$ ($\forall i \in V$).

Each follower updates its current position according to Eq. (3.21), where $g_i = g = 30$ ($\forall i \in V_F$), and the communication weights are consistent with the initial positions shown in Fig. 3.1 as listed in Table 3.1. The X and Y components of the desired and actual positions of the follower 18 are depicted versus time in Fig. 3.8. As it is seen in Fig. 3.8, follower 18 deviates from its desired position during transition, however, it ultimately reaches the final desired position, $(X_{F,18}, Y_{F,18}) = (3.3437, 12.7074)$.

Parameters $p_{18,1}, p_{18,2}$, and $p_{18,3}$, calculated by using Eq. (2.13) based on the final positions of the follower 18 and the, are as follows:

$$
\begin{bmatrix} p_{18,1} \\ p_{18,2} \\ p_{18,3} \end{bmatrix} = \begin{bmatrix} X_{1,F} & X_{2,F} & X_{3,F} \\ Y_{1,F} & Y_{2,F} & Y_{3,F} \\ 1 & 1 & 1 \end{bmatrix}^{-1} \begin{bmatrix} X_{18,F} \\ Y_{18,F} \\ 1 \end{bmatrix} = \begin{bmatrix} 6 & 0 & 6 \\ 11.5 & 11.5 & 16 \\ 1 & 1 & 1 \end{bmatrix}^{-1} \begin{bmatrix} 3.3437 \\ 12.7074 \\ 1 \end{bmatrix} = \begin{bmatrix} 0.2890 \\ 0.4427 \\ 0.2683 \end{bmatrix}.
$$

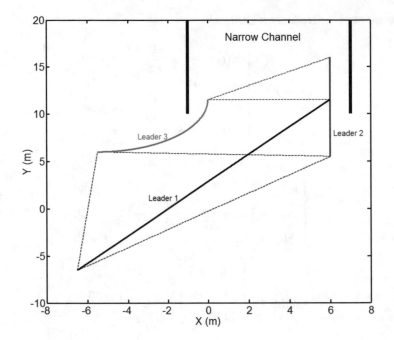

Fig. 3.6 Paths of the leaders in example 3.2

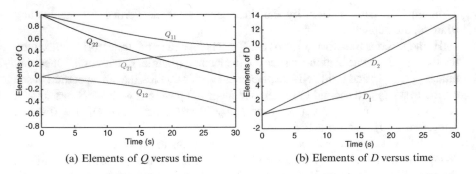

(a) Elements of Q versus time

(b) Elements of D versus time

Fig. 3.7 Entries Q and D versus time

As it is observed, the ultimate values of $p_{18,1}$, $p_{18,2}$, $p_{18,3}$ are the same as $\alpha_{18,1}$, $\alpha_{18,2}$, and $\alpha_{18,3}$ which are uniquely determined based on the initial positions of the follower 18 and the three leaders as follows:

$$
\begin{bmatrix} \alpha_{18,1} \\ \alpha_{18,2} \\ \alpha_{18,3} \end{bmatrix} = \begin{bmatrix} X_1 & X_2 & X_3 \\ Y_1 & Y_2 & Y_3 \\ 1 & 1 & 1 \end{bmatrix}^{-1} \begin{bmatrix} X_{18} \\ Y_{18} \\ 1 \end{bmatrix} = \begin{bmatrix} -6.5 & -5.5 & 6 \\ -6.5 & 6 & 5 \\ 1 & 1 & 1 \end{bmatrix}^{-1} \begin{bmatrix} -2.6927 \\ 2.1270 \\ 1 \end{bmatrix} = \begin{bmatrix} 0.2883 \\ 0.4425 \\ 0.2692 \end{bmatrix}.
$$

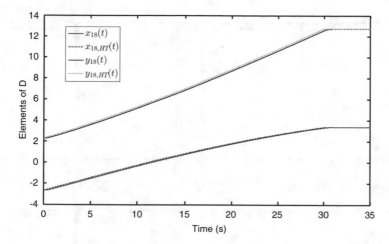

Fig. 3.8 X and Y components of the desired and actual positions of the follower 18

This implies that final formation of the MAS is a homogeneous transformation of the initial configuration.

Formations of the MAS at six sample times $t = 5s$, $t = 10s$, $t = 15s$, $t = 20s$, $t = 25s$, and $t = 35s$ are shown in Fig. 3.9.

Followers with Finite Body Size

As observed above, followers deviate from the desired positions defined by a homogeneous deformation during transition, when position of the follower $i \in V_F$ is updated according to the first order dynamics (3.21). In this section, it is desired to specify an upper bound for the deviations of followers. Therefore, avoidance of interagent collision can be assured, when each follower has a finite size.

Constraints on leaders' motion: The following four constraints are considered for motion of the leaders:

- Positions of leaders are required to satisfy the rank condition (2.3) at any time t. This condition assures that eigenvalues of the Jacobian matrix $Q(t)$ remain positive at any time t, and thus the deformation mapping remains nonsingular.
- Paths of the leaders are chosen such that the leading polytope does not collide with obstacles in the motion field. A schematic of path planning, illustrating avoidance of collision of the leading triangle in the motion plane, is shown in Fig. 3.10.
- It is required that $n+1$ leaders choose their trajectories such that desired positions of no two followers get closer than $2(\delta + \varepsilon)$,

$$i, j \in V \wedge i \neq j, \|r_{i,HT}(t) - r_{j,HT}(t)\| \geq 2(\delta + \varepsilon). \tag{3.28}$$

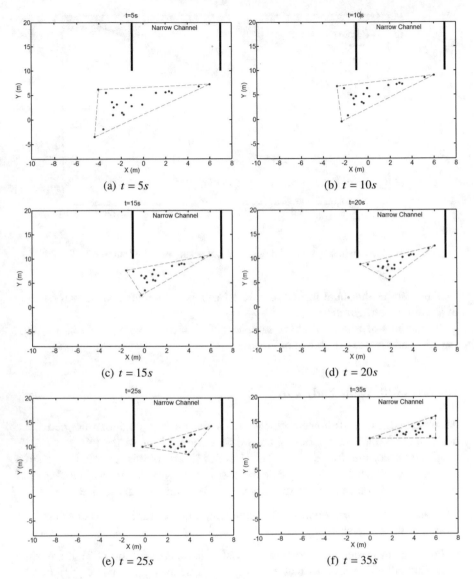

Fig. 3.9 MAS formations at five different sample times $t = 5s$, $t = 10s$, $t = 15s$, $t = 20s$, $t = 25s$, and $t = 35s$

Note that

$$\delta \geq ||r_i(t) - r_{i,HT}(t)|| \tag{3.29}$$

is the upper limit for deviation of follower agent $i \in V_F$ form the desired position $r_{i,HT}(t)$ given by a homogeneous deformation (δ is obtained in the sequel.). When each follower acquires homogeneous deformation through local communication.

Fig. 3.10 Schematic of paths
chosen by leaders assuring
avoidance of agents with
obstacle

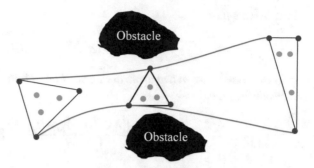

Fig. 3.11 Graphical
representations of γ_l and γ_s

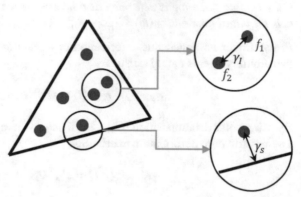

- It is also necessary that distance of each follower from boundary of the leading
  polytope remains equal or greater than $(\delta + \varepsilon)$. This requirement guarantees that
  no follower leaves the leading polytope.

It is supposed that $f_1 \in V$ and $f_2 \in V$ are index numbers of two agents that have
minimum separation distance

$$\gamma_l = ||R_{f_1} - R_{f_2}|| \tag{3.30}$$

in the initial formation; γ_s is the minimum distance of a follower from the
boundary of the leading polytope in the initial formation of the MAS. Graphical
representations of γ_l and γ_s are shown in Fig. 3.11 for a typical initial distribution of
agents in the $X - Y$ plane.

Theorem 3.2. *Suppose that $\lambda_{min} \in \mathbb{R}_+$ is the lower limit for the magnitudes of
eigenvalues of the matrix $\sqrt{Q^T Q} \in \mathbb{R}^{n \times n}$,*

$$\forall t \geq t_0, \ \lambda_{min} \leq \lambda_1(\sqrt{Q^T Q}) < \cdots < \lambda_n(\sqrt{Q^T Q}).$$

Also, assume that

$$\gamma_{ij}(t) = ||r_{i,HT}(t) - r_{j,HT}(t)||$$

is the distance between two arbitrary agents i and j initially positioned at R_i and R_j, respectively. Then

$$\frac{\gamma_{ij}(t)}{\gamma_{ij}(t_0)} = \frac{||r_{i,HT}(t) - r_{j,HT}(t)||}{||R_i - R_j||} \geq \lambda_{min}, \tag{3.31}$$

if the leading polytope is deformed as a homogeneous transformation. Notice that t_0 is the initial time and $\gamma_{ij}(t_0) = ||R_i - R_j||$.

Proof. Under homogeneous deformation of a leading polytope, distance between two arbitrary agents i and j satisfies

$$r_{i,HT}(t) - r_{j,HT}(t) = Q(t)(R_i - R_j). \tag{3.32}$$

By using polar decomposition [64], $Q = R_O U_D$, where R_O is orthogonal and U_D is a symmetric pure deformation matrix. So,

$$Q^T Q = U_D{}^T U_D \tag{3.33}$$

and

$$\lambda(U_D) = \lambda(\sqrt{Q^T Q}). \tag{3.34}$$

By considering definition of the homogeneous transformation,

$$r_{i,HT}(t) - r_{j,HT}(t) = Q(t)(R_i - R_j), \tag{3.35}$$

therefore,

$$(r_{i,HT}(t) - r_{j,HT}(t))^T (r_{i,HT}(t) - r_{j,HT}(t)) = (R_i - R_j)^T Q^T Q(R_i - R_j) \geq$$
$$(R_i - R_j)^T \lambda_{min}{}^2 (R_i - R_j). \tag{3.36}$$

Consequently,

$$\lambda_{min} \leq \frac{||r_{i,HT}(t) - r_{j,HT}(t)||}{||R_i - R_j||} = \frac{\gamma_{ij}(t)}{\gamma_{ij}(t_0)}. \tag{3.37}$$

Remark 3.4. Given ε (the radius of each follower), γ_l (initial distance between the agents $f_1 \in V$ and $f_2 \in V$ that are at the closest distance in the initial configuration) and γ_s (smallest distance of an agent from the boundary of the leading polytope), then

$$\delta_{max} = min\{\frac{1}{2}(\gamma_l - 2\varepsilon), (\gamma_s - \varepsilon)\}. \tag{3.38}$$

Note that no two followers get closer than $2(\delta + \varepsilon)$ and no follower leaves the leading polytope, if

$$\lambda_{min} \leq \frac{2(\delta + \varepsilon)}{2(\delta_{max} + \varepsilon)}. \tag{3.39}$$

It is noticed that δ is the upper limit for deviations of followers from the state of homogeneous transformation, when followers learn homogeneous deformation by local communication. Because entries of Q and D can be uniquely related to the components of leaders' positions by using Eq. (2.8), it is necessary that trajectories of leaders are specified such that the minimum value of the smallest eigenvalues of the matrix $\sqrt{Q^T Q}$ never becomes less than λ_{min} given by Eq. (3.39).

Upper Bound δ for Followers' Deviations: Let dynamics of evolution of followers be expressed by

$$\frac{dz_q}{dt} = g(Az_q + Bu_q) = gA(z_q - z_{q,HT}) = gAE_q, \tag{3.40}$$

where $E_q = z_{q,HT} - z_q$ is the transient error which is updated by

$$\frac{dE_q}{dt} - gAE_q = \dot{z}_{q,HT}. \tag{3.41}$$

Then

$$E_q(t) = e^{gAt}E_q(t_0) + \int_{t_0}^t e^{gA(t-\tau)}\dot{z}_{q,HT}d\tau. \tag{3.42}$$

Because initial positions of the agents satisfy Eq. (3.10),

$$z_q(t_0) = -A^{-1}Bu_q(t_0),$$

$z_{q,HT}(t_0) = z_q(t_0) = -A^{-1}Bu_q(t_0)$ and $E_q(t_0) = z_{q,HT}(t_0) - z_q(t_0)$ vanishes. Therefore,

$$E_q(t) = \int_{t_0}^t e^{gA(t-\tau)}\dot{z}_{q,HT}d\tau. \tag{3.43}$$

Lemma. Let $\lambda_1, \lambda_2, \ldots, \lambda_{N-n-1}$ be the eigenvalues of the matrix A, then

$$\forall i \in \{1, \ldots, N-n-1\}, \frac{1}{||A^{-1}||} \leq \lambda_i \leq ||A||. \tag{3.44}$$

Theorem 3.3. *Let $V \in \mathbb{R}_+$ be the upper limit for the magnitudes of the velocities of the $n + 1$ leaders, where dynamics of evolution of followers are given by Eq. (3.40). Then*

$$\delta = \frac{\|A^{-1}\|\sqrt{N-n-1}}{g} V \tag{3.45}$$

specifies an upper bound for deviation of each follower from the desired position $r_{i,HT}(t)$ given by a homogeneous transformation, where N is the total number of agents (leaders and followers) of the MAS.

Proof. $E_q(t)$ determined by Eq. (3.42) satisfies the following inequality:

$$\|E_q(t)\| = \|\int_{t_0}^{t} e^{gA(t-\tau)} \dot{z}_{q,HT} d\tau\| \leq \|\dot{z}_{q,HT}\| \, \|\int_{t_0}^{t} e^{gA(t-\tau)} d\tau\|. \tag{3.46}$$

By considering the above Lemma, it is concluded that

$$\|E_q(t)\| = \|\int_{t_0}^{t} e^{gA(t-\tau)} \dot{z}_{q,HT} d\tau\| \leq \|\dot{z}_{q,HT}\| \, \|\int_{t_0}^{t} e^{gA(t-\tau)} d\tau\| \leq \frac{\|\dot{z}_{q,HT}\| \, \|A^{-1}\|}{g}. \tag{3.47}$$

It is assumed that V_q is the upper bound for the qth components of the velocities of the leaders. Therefore,

$$\|\dot{z}_{q,HT}\| = \|-A^{-1}B\dot{u}_q\| = \|W_L\dot{u}_q\| \leq V_q\|W_L\mathbf{1}\| \tag{3.48}$$

where $\mathbf{1} \in \mathbb{R}^{n+1}$ is the one vector, and the positive matrix $W_L \in \mathbb{R}^{(N-n-1)\times n+1}$ is one-sum-row (sum of each row of W_L is equal to 1.). This implies that

$$\|W_L\mathbf{1}\| = \|\mathbf{1}\| = \sqrt{N-n-1}, \tag{3.49}$$

$$\|\dot{z}_{q,HT}\| \leq V_q\sqrt{N-n-1}, \tag{3.50}$$

and

$$\|E_q(t)\| \leq \delta_q = \frac{\|A^{-1}\|\sqrt{N-n-1}}{g} V_q. \tag{3.51}$$

By considering the inequality (3.51), it is concluded that

$$\|x_q(t) - x_{q,HT}(t)\| \leq \|E_q(t)\| \leq \delta_q \tag{3.52}$$

where

$$\delta_q = \frac{||A^{-1}||\sqrt{N-n-1}}{g} V_q.$$ (3.53)

Let

$$V = \sqrt{\sum_{q=1}^{n} V_q^2}$$

be the upper limit for the velocities of the leaders at any time t during MAS evolution, then

$$\delta = \sqrt{\sum_{q=1}^{n} \delta_q^2} = \frac{||A^{-1}||\sqrt{N-n-1}}{g} V$$ (3.54)

specifies an upper bound for deviation of each follower from the desired state defined by a homogeneous transformation. In other words,

$$||r_i(t) - r_{i,HT}(t)|| \leq \delta = \frac{||A^{-1}||\sqrt{N-n-1}}{g} V.$$ (3.55)

Example 3.3. Consider an MAS that contains 10 agents (3 leaders and 7 followers) with the initial positions shown in Fig. 3.4. The MAS is supposed to negotiate the narrow passage shown in Fig. 3.12. For this purpose, leaders choose the paths shown in Fig. 3.12, where X and Y components of positions of the leaders are given by

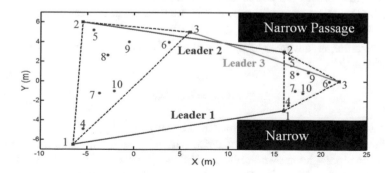

Fig. 3.12 Paths of the Leaders in Example 3.3

$$
Leader1 : \begin{cases} x_1(t) = \dfrac{22.5}{T}t - 6.5 & 0 \le t \le T \\[2mm] x_1(t) = 16 & t > T \\[2mm] y_1(t) = \dfrac{3.5}{T}t - 6.5 & 0 \le t \le T \\[2mm] y_1(t) = -3 & t > T \end{cases}
$$

$$(3.56)$$

$$
Leader2 : \begin{cases} x_2(t) = \dfrac{21.5}{T}t - 5.5 & 0 \le t \le T \\[2mm] x_2(t) = 16 & t > T \\[2mm] y_2(t) = \dfrac{-3}{T}t + 6 & 0 \le t \le T \\[2mm] y_2(t) = 3 & t > T \end{cases}
$$

$$
Leader3 : \begin{cases} x_3(t) = \dfrac{16}{T}t + 6 & 0 \le t \le T \\[2mm] x_3(t) = 22 & t > T \\[2mm] y_3(t) = \dfrac{-5}{T}t + 5 & 0 \le t \le T \\[2mm] y_3(t) = 0 & t > T \end{cases} .
$$

Given positions of leaders, magnitudes of the velocities of the leaders do not exceed

$$
V = \frac{\sqrt{22.5^2 + 3.5^2}}{T} = \frac{22.7706}{T}. \tag{3.57}
$$

Shown in Fig. 3.13 are entries of the Jacobian Q (Q_{11}, Q_{12}, Q_{21}, Q_{22}) and the vector D (D_1, D_2) versus t/T, where T is the time when leaders reach the desired final positions at $(16, -3)$, $(16, 3)$, and $(22, 0)$.

Furthermore, eigenvalues of the matrix $\sqrt{Q^T Q}$ are shown in Fig. 3.14. As it is observed,

$$
\lambda_{min} = 0.3914 \tag{3.58}
$$

is the lower bound for the eigenvalues of the matrix $\sqrt{Q^T Q}$.

Followers acquire homogeneous deformation by applying three different communication topologies shown in Fig. 3.4 with weight matrix $W_1 = [A_1 \ B_1]$, Fig. 3.15 with weight matrix $W_2 = [A_2 \ B_2]$, and Fig. 3.16 with weight matrix $W_3 = [A_3 \ B_3]$. Given communication weights listed in Tables 3.2, 3.4, and 3.5, the matrices $A_1, B_1, A_2, B_2, A_3,$ and B_3 become

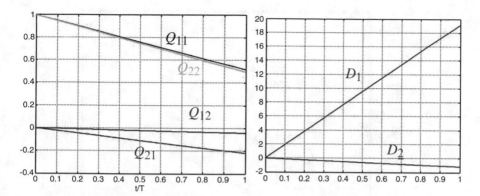

Fig. 3.13 Entries of the Jacobian matrix Q and the vector D

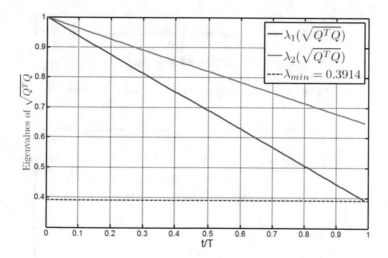

Fig. 3.14 Eigenvalues of the Jacobian matrix Q

$$
A_1 = \begin{bmatrix}
-1 & 0 & 0 & 0.15 & 0 & 0 & 0.15 \\
0 & -1 & 0 & 0 & 0.15 & 0.15 & 0 \\
0 & 0 & -1 & 0 & 0 & 0.15 & 0.15 \\
0.40 & 0 & 0 & -1 & 0.36 & 0 & 0.24 \\
0 & \dfrac{1}{3} & 0 & \dfrac{1}{3} & -1 & \dfrac{1}{3} & 0 \\
0 & 0.31 & 0.42 & 0 & 0.27 & -1 & 0 \\
0.35 & 0 & 0.29 & 0.36 & 0 & 0 & -1
\end{bmatrix}
$$

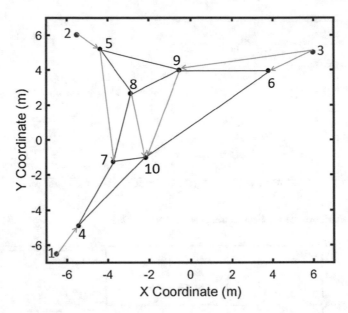

Fig. 3.15 A typical graph showing non-minimum directed communication

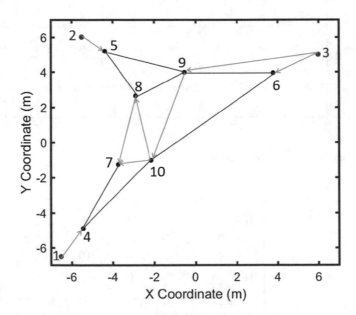

Fig. 3.16 A typical graph showing minimum directed communication

$$B_1 = \begin{bmatrix} 0.7 & 0 & 0 \\ 0 & 0.7 & 0 \\ 0 & 0 & 0.7 \\ 0 & 0 & 0 \\ 0 & 0 & 0 \\ 0 & 0 & 0 \\ 0 & 0 & 0 \end{bmatrix}$$

$$A_2 = \begin{bmatrix} -1 & 0 & 0 & 0.15 & 0 & 0 & 0.15 \\ 0 & -1 & 0 & 0 & 0.15 & 0.15 & 0 \\ 0 & 0 & -1 & 0 & 0.1300 & 0.0067 & 0.1234 \\ 0.3600 & 0.1661 & 0 & -1 & 0.0357 & 0 & 0.4832 \\ 0 & 0.4700 & 0.1435 & 0.3641 & -1 & 0.0224 & 0.3777 \\ 0 & 0.2500 & 0.0440 & 0 & 0.4333 & -1 & 0 \\ 0.5600 & 0 & 0.0007 & 0.0007 & 0.0128 & 0.1566 & -1 \end{bmatrix}$$

$$B_2 = \begin{bmatrix} 0.7 & 0 & 0 \\ 0 & 0.7 & 0 \\ 0 & 0 & 0.74 \\ 0 & 0 & 0 \\ 0 & 0 & 0 \\ 0 & 0 & 0.2726 \\ 0 & 0 & 0 \end{bmatrix}$$

$$A_3 = \begin{bmatrix} -1 & 0 & 0 & 0.15 & 0 & 0 & 0.15 \\ 0 & -1 & 0 & 0 & 0.15 & 0.15 & 0 \\ 0 & 0 & -1 & 0 & 0 & 0.15 & 0.15 \\ 0.40 & 0 & 0 & -1 & 0.36 & 0 & 0.24 \\ 0 & 0.4523 & 0 & 1/3 & -1 & 0.1700 & 0.3777 \\ 0 & 0.31 & 0.42 & 0 & 0.27 & -1 & 0 \\ 0.5623 & 0 & 0.2653 & 0.36 & 0 & 0.1724 & -1 \end{bmatrix}$$

$$B_3 = \begin{bmatrix} 0.7 & 0 & 0 \\ 0 & 0.7 & 0 \\ 0 & 0 & 0.7 \\ 0 & 0 & 0 \\ 0 & 0 & 0 \\ 0 & 0 & 0 \\ 0 & 0 & 0 \end{bmatrix}.$$

Table 3.4 Communication weights - graph shown in Fig. 3.15

Index i	i_1	i_2	i_3	i_4	i_5	w_{i,i_1}	w_{i,i_2}	w_{i,i_3}	w_{i,i_4}	w_{i,i_5}
4	1	7	8	–	–	0.70	0.15	0.15	–	–
5	2	8	9	–	–	0.70	0.15	0.15	-	–
6	3	9	10	8	–	0.74	0.01	0.12	0.13	–
7	4	8	4	5	–	0.36	0.04	0.44	0.16	–
8	5	7	9	6	–	0.47	0.36	0.03	0.14	–
9	5	6	8	3	–	0.30	0.36	0.29	0.05	–
10	4	6	7	8	9	0.56	0.27	0.01	0.01	0.16

Table 3.5 Communication weights - graph shown in Fig. 3.16

Index i	i_1	i_2	i_3	w_{i,i_1}	w_{i,i_2}	w_{i,i_3}
4	1	7	10	0.70	0.15	0.15
5	2	8	9	0.70	0.15	0.15
6	3	9	10	0.70	0.15	0.15
7	4	8	10	0.40	0.36	0.24
8	5	9	10	0.45	0.17	0.38
9	5	6	8	0.31	0.42	0.27
10	4	6	9	0.56	0.27	0.17

Table 3.6 Eigenvalues of the matrix A_1

	λ_1	λ_2	λ_3	λ_4	λ_5	λ_6	λ_7
Re	−0.29	−0.59	−0.83	−1.11	−1.21	−1.43	−1.53
Im	0	0	0	0	0	0	0

It is noticed that

$$W_L = -A_1^{-1}B_1 = -A_2^{-1}B_2 = -A_3^{-1}B_3 = \begin{bmatrix} 0.8647 & 0.0538 & 0.0815 \\ 0.0562 & 0.8397 & 0.1040 \\ 0.0991 & 0.0836 & 0.8173 \\ 0.5636 & 0.2323 & 0.2041 \\ 0.2487 & 0.5009 & 0.2504 \\ 0.1262 & 0.4307 & 0.4431 \\ 0.5343 & 0.1267 & 0.3390 \end{bmatrix}.$$

Also,

$$\|A_1^{-1}\| = 3.5775$$

$$\|A_2^{-1}\| = 3.1502.$$

$$\|A_3^{-1}\| = 2.9106$$

Furthermore, eigenvalues of the matrices A_1, A_2, and A_3 are given in Tables 3.6, 3.7, and 3.8, respectively. As it is observed $\lambda_s(A_1) < \lambda_s(A_3) < \lambda_s(A_2)$ e.g. λ_s denotes the smallest eigenvalue.

Table 3.7 Eigenvalues of the matrix A_2

	λ_1	λ_2	λ_3	λ_4	λ_5	λ_6	λ_7
Re	−0.372	−0.70	−0.97	−1.15	−1.15	−1.27	−1.39
Im	0	0	0	0.24	−0.24	0	0

Table 3.8 Eigenvalues of the matrix A_3

	λ_1	λ_2	λ_3	λ_4	λ_5	λ_6	λ_7
Re	−0.36	−0.66	−0.88	−1.16	−1.16	−1.35	−1.43
Im	0	0	0	0.21	−0.21	0	0

Assurance of Collision Avoidance: Given initial positions of the agents,

$$\gamma_l = ||R_7 - R_{10}|| = \sqrt{(-3.7164 - (-2.1356))^2 + (-1.2494 - (-1.0174))^2} = 1.5977.$$

Additionally, the follower 4 has the minimum distance

$$\gamma_s = \frac{|(X_3 - X_1)(Y_1 - Y_4) - (Y_3 - Y_1)(X_1 - X_4)|}{\sqrt{(X_3 - X_1)^2 + (Y_3 - Y_1)^2}} = 0.4589$$

from the boundaries of the leading polytope at $t = 0$. It is assumed that each follower is a disk with radius $\varepsilon = 0.04m$, therefore,

$$\delta_{max} = min\{\frac{1}{2}(\gamma_l - 2\varepsilon), (\gamma_s - \varepsilon)\} = 0.4189.$$

Let

$$\lambda_{min} = \frac{\delta + \varepsilon}{\delta_{max} + \varepsilon} = 0.3914,$$

then

$$\delta = 0.1393,$$

is the upper limit for the deviation of each follower from the desired state given by a homogeneous deformation. In other words, avoidance of inter-agent collision is assured, if $||r_i(t) - r_{i,HT}(t)|| \le \delta = 0.1393$.

On the other hand, substituting $||A_1^{-1}|| = 3.5775$, $N = 10$, $n = 2$, and $V = \dfrac{\sqrt{22.5^2 + 3.5^2}}{T}$ into Eq. (3.55) results in

$$\frac{3.5775 \times \sqrt{7} \times 22.7706}{gT} V \le 0.1393$$

or

$$gT \ge 1547.2. \tag{3.59}$$

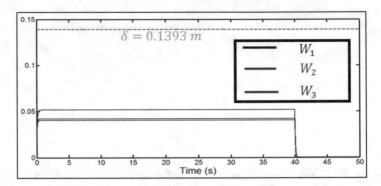

Fig. 3.17 Deviation of follower 10 from desired position given by a homogeneous transformation ($\|r_{10,HT} - r_{10}\|$) under three different communication topologies $W_1 = [A_1 \ B_1]$, $W_2 = [A_2 \ B_2]$, and $W_3 = [A_3 \ B_3]$

The control gain $g = 40$ is applied by each follower, where leaders reach their final positions at $T = 40s$. Thus, the inequality (3.59) is satisfied.

Deviation of the follower 10 from the desired position ($\|r_{10,HT} - r_{10}\|$) is shown versus time in Fig. 3.17 by black and blue and red curves, where followers apply three different communication topologies $W_1 = [A_1 \ B_1]$, $W_2 = [A_2 \ B_2]$, and $W_3 = [A_3 \ B_3]$, respectively. As it is observed, deviation of follower 10 does not exceed $\delta = 0.1393$ during MAS evolution; therefore, it can be assured that both interagent collision and collisions of agents with obstacles are avoided. It is also seen that deviation of follower 10 is the most, when topology W_1 associated with communication graph shown in Fig. 3.4 is applied. Also, the least deviation is observed when the topology W_2 with non-minimum directed communication graph shown in Fig. 3.15 is applied. This is because $\lambda_s(A_1) < \lambda_s(A_3) < \lambda_s(A_2)$ and thus followers have the least deviations from the state of homogeneous deformation, when they apply the communication graph shown in Fig. 3.15.

3.2.3.2 Dynamics of Followers in Presence of Communication Delays

Let position of the follower $i \in V_F$ be updated by Eqs. (3.18) and (3.19), where the follower i applies the control gain $g_i = g$ ($\forall i \in V_F$), and the time delays $h_1 = h_{1i}$ and $h_2 = h_{2i}$ are not both zero simultaneously. Then, the qth components of followers' positions are updated by the following first order dynamics:

$$\dot{z}_q(t) = -gz_q(t - h_1) + gFz_q(t - h_2) + \beta F\dot{z}_q(t - h_2) + gBu_q(t - h_2) + \beta B\dot{u}_q(t - h_2). \tag{3.60}$$

It is noted that $F = A + I$ is an irreducible and nonnegative matrix with the spectral radius $\rho(F)$ that is less than 1. The characteristic equation of the first order dynamics (3.60) becomes

$$CHE(s, h_1, h_2) = |sI + gIe^{-h_1 s} - gFe^{-h_2 s} - \beta sFe^{-h_2 s}| = 0. \tag{3.61}$$

MAS Evolution without Self-Delay ($h_1 = 0$): The characteristic Eq. (3.61) is stable, if $h_1 = 0$ and $h_2 > 0$. The proof is provided in the theorem below.

Theorem 3.4. *If the follower $i \in V_F$ perceives its own position without time delay ($h_1 = 0$), then the communication delay $h_2 = h \geq 0$ in perceiving positions of the in-neighbor agents does not affect stability of evolution of followers.*

Proof. If $h_1 = 0$ and $h_2 = h$, then the characteristic Eq. (3.61) simplifies to

$$CHE(s,h) = |(s+g)I - (g+\beta s)Fe^{-hs}| = 0. \tag{3.62}$$

Notice that the roots of the characteristic equation (3.62) is the same as the roots of the following characteristic equation:

$$CHE(s,h) = |e^{hs}\frac{s+g}{\beta s+g}I - F| = 0 \tag{3.63}$$

when $s = j\omega$ is substituted in both Eqs. (3.62) and (3.63).

Because $\rho(F) < 1$, stability of MAS evolution is assured, if

$$|\sigma(s)| = |\frac{e^{hs}(s+g)}{\beta s+g}| < 1. \tag{3.64}$$

If $\beta = 1$, then $\sigma(s) = e^{hs}$. Therefore, the condition $|\sigma(s)| = |e^{hs}| < 1$ is satisfied for any time delay $h > 0$, when real part of s is negative. In other words, the roots of the characteristic equation (3.61) are all placed in the open left half s-plane for any communication delay $h > 0$.

If $\beta = 0$, then Eq. (3.64) simplifies to

$$|e^{hs}(s+g)| < g. \tag{3.65}$$

Substituting $s = x + iy$ in the inequality (3.65) and then squaring both sides leads to

$$e^{hx}((x+g)^2 + y^2) - g^2 < 0. \tag{3.66}$$

The inequality (3.66) is satisfied only when $x < 0$. Therefore, the communication delay $h > 0$ does not influence stability of MAS evolution.

MAS Evolution with Self-Delay ($h_1 = h_2 = h$): For this case, the characteristic equation (3.61) simplifies to

$$CHE(s,h_1,h_2) = |sI - gAe^{-hs} - \beta sFe^{-hs}| = 0. \tag{3.67}$$

To assure stability, all roots of the characteristic Eq. (3.67) must be located in the open left half s-plane. However, because of transcendental term e^{-hs}, Eq. (3.67) has an infinite number of roots and it is difficult to check, under what condition, its roots are located in the open left half s-plane. For the stability analysis, a method called *cluster treatment of characteristic roots* (CTCR) [94, 131] is applied.

If the MAS evolution dynamics is to go unstable, when h increases to its maximum allowable value (dented by h_{all}), some roots of the characteristic Eq. (3.67) cross the $j\omega$ axis from the left half of the s-plane into the right half s-plane. Thus, the stability of the time delay system can be specified by looking for solutions of the characteristic Eq. (3.67) when s is substituted by $j\omega$. This may be handled by the first order Pade approximation,

$$e^{-jh\omega} = \frac{1 - jT\omega}{1 + jT\omega}, \tag{3.68}$$

where

$$h = \frac{2}{\omega} \tan^{-1}(\omega T + k\pi), \; k \in \mathbb{Z}. \tag{3.69}$$

Consequently instead of studying $CHE(j\omega, h_1, h_2)$ in Eq. (3.67), roots of

$$\left| sI - g\frac{1 - Ts}{1 + Ts}A - \beta s\frac{1 - Ts}{1 + Ts}F \right| = 0 \tag{3.70}$$

is checked when $s = j\omega$ is substituted in Eq. (3.70).

The roots of Eq. (3.70) are the same as the roots of the following characteristic equation:

$$\left| s^2 T(I + F) + s((I - \beta F) - Tg(I - F)) + g(I - F) \right| = 0. \tag{3.71}$$

Equation (3.72) can be rewritten as

$$\sum_{k=0}^{2(N-n-1)} \tau_k(T)s^k = 0. \tag{3.72}$$

Therefore, the stability of the system can be ascertained by using the Routh's stability criterion to estimate the maximum value for T, T_{all}, if one exists.

Remark 3.5. Because the MAS evolution dynamics without time delay is stable, from continuity, the dynamics of the delayed system will remain stable for $0 \leq h < h_{all}$.

Remark 3.6. For $T = T_{all}$, the characteristic Eq. (3.61) has possibly finite number of imaginary roots at $j\omega_1, j\omega_2, \ldots, j\omega_m$. Therefore, m clusters of time delays

$$h_{ik} = \frac{2}{\omega_i} \tan^{-1}(\omega_i T_{all} + k\pi), \; k \in \mathbb{Z}, \; i = 1, 2, \ldots, m \tag{3.73}$$

are obtained, where every cluster h_{ik} has an infinite number of members. In other words, for every ω_i ($i = 1, 2, \ldots, m$) and given T_{all}, an infinite number of h_{ik} can be obtained. The smallest positive value of h_{ik} is considered as the maximum allowable time delay, denoted by h_{all}. Let $\omega_1 \leq \omega_2 \leq \cdots \leq \omega_m$ at $T = T_{all}$, then

$$h_{all} = \frac{2}{\omega_m} \tan^{-1}(\omega_m T_{all}). \tag{3.74}$$

This is concluded from the fact that h_{ik} in Eq. (3.73) is a decreasing function of ω_i.

Stability Analysis of MAS evolution when $\beta = 1$ *and* $h_1 = h_2 = h$ *and eigenvalues of F are real:* Under this scenario, the characteristic Eq. (3.71) simplifies to

$$|s^2 T(I+F) + s(I-F)(1-Tg) + g(I-F)| = 0. \tag{3.75}$$

The roots of the characteristic Eq. (3.75) are the same as the roots of the following characteristic equation:

$$\left| \frac{s^2 T}{s(1-Tg)+g} I - [-(I+F)^{-1}(I-F)] \right| = 0. \tag{3.76}$$

Since $\rho(F)$ (the spectrum of F) is less than 1, eigenvalues of the matrices $I - F$ and $I + F$ are located in the open right half s-plane. Therefore, eigenvalues of the matrix $[-(I+F)^{-1}(I-F)]$ are located in the open right half s-plane. If eigenvalues of the matrix F are all real, then eigenvalues of the matrix $[-(I+F)^{-1}(I-F)]$ are real and negative. In other words,

$$Re\{\frac{s^2}{s(1-Tg)+g}\} < 0, \tag{3.77}$$

$$Im\{\frac{s^2}{s(1-Tg)+g}\} = 0. \tag{3.78}$$

Replacing $s = x + jy$ in Eqs. (3.77) and (3.78) leads to the following condition:

$$y^2 = -x^2 - \frac{2g}{1-Tg} x \geq 0. \tag{3.79}$$

By substituting $y = 0$ in Eq. (3.79)

$$x \leq -\frac{2g}{1-Tg}. \tag{3.80}$$

Because $x \leq 0$ inequality (3.80) is satisfied when $1 - Tg$ is positive. On the other hand $t > 0$, because $h > 0$. Thus, $0 \leq T < T_{all} = \frac{1}{g}$. If $T = \frac{1}{g}$ is substituted in

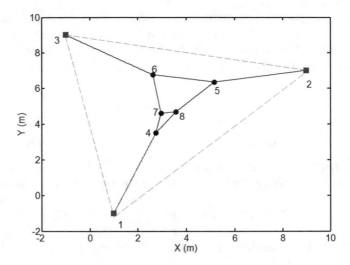

Fig. 3.18 Communication graph applied by followers in Example 3.4 to acquire the desired poisons defined by a homogeneous transformation

Table 3.9 Initial positions and communication weights associated with the formation shown in Figure 3.18

i	X_i	Y_i	i_1	i_2	i_3	w_{i,i_1}	w_{i,i_2}	w_{i,i_3}
1	1.0000	5.000	–	–	–	–	–	–
2	3.0000	7.0000	–	–	–	–	–	–
3	5.0000	8.0000	–	–	–	–	–	–
4	2.4695	6.2405	1	7	8	0.20	0.50	0.30
5	3.2365	6.9702	2	6	8	0.36	0.40	0.24
6	3.7048	7.1947	3	5	7	0.32	0.43	0.25
7	2.8526	6.5501	4	6	8	0.48	0.23	0.29
8	2.8107	6.5514	4	5	7	0.43	0.32	0.25

the characteristic Eq. (3.76), then imaginary roots (of Eq. (3.76)) are obtained as follows:

$$j\omega_i = \sqrt{\eta_i},\tag{3.81}$$

where η_i is the ith ($i = 1, 2,\ldots, N - n - 1$) eigenvalue of the matrix $[-g^2(I + F)^{-1}(I - F)]$.

Example 3.4 ([113]). Consider an MAS consisting of 8 agents with three leaders and five followers. Initial positions of the leaders and followers are listed in Table 3.9. Followers apply the graph shown in Figure 3.18 to acquire the desired positions by local communication. Followers' communication weights are listed in Table 3.9, where they are consistent with the agents' initial positions and the graph shown in Fig. 3.18. By using the definition (3.8), the weight matrix $W \in \mathbb{R}^{5\times5}$ is obtained as follows:

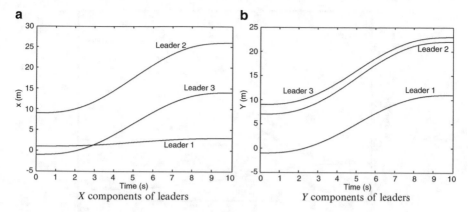

Fig. 3.19 X and Y components of leaders' positions (**a**) X components of leaders' positions (**b**) Y components of leaders' positions

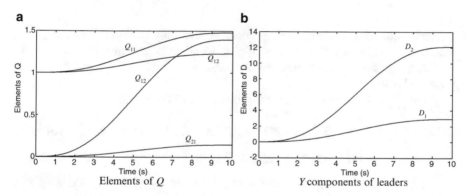

Fig. 3.20 Elements of Q and D versus time

$$W = \begin{bmatrix} 0.20 & 0 & 0 & -1 & 0 & 0 & 0.50 & 0.30 \\ 0 & 0.36 & 0 & 0 & -1 & 0.40 & 0 & 0.24 \\ 0 & 0 & 0.32 & 0 & 0.43 & -1 & 0.25 & 0 \\ 0 & 0 & 0 & 0.48 & 0 & 0.23 & -1 & 0.29 \\ 0 & 0 & 0 & 0.43 & 0.32 & 0 & 0.25 & -1 \end{bmatrix}. \tag{3.82}$$

Eigenvalues of the matrix $F = A + I$ are $\lambda_{1F} = 0.8522$, $\lambda_{2F} = 0.3511$, $\lambda_{3F} = -0.0765$, $\lambda_{4F} = -0.5116$, and $\lambda_{5F} = -0.6152$. Therefore, the matrix $A = -(I - F)$ is Hurwitz because eigenvalues of F are all located inside the unit disk centered at the origin.

Components of leaders' positions in the $X - Y$ plane are shown in Fig. 3.19. Given leaders' positions, elements of Q and D are obtained by using Eq. (2.21) as shown in Fig. 3.20. Furthermore, eigenvalues of the pure deformation matrix $U_D = \sqrt{Q^T Q}$ are shown in Fig. 3.21.

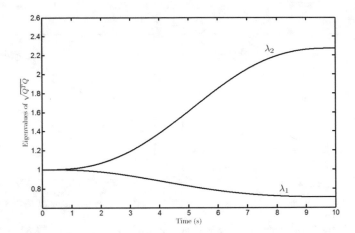

Fig. 3.21 Eigenvalues of the matrix $\sqrt{Q^T Q}$ versus time

Each follower updates its current position according to Eqs. (3.18) and (3.19), where $h_{1_i} = h_{2_i} = h$ and $g_i = g = 5(\forall i \in V_F)$. The simulation results of MAS evolution for the two scenarios that are associated with $\beta = 0$ and $\beta = 1$ are presented below.

$\beta = 0$: For this scenario, the characteristic Eq. (3.71) simplifies to

$$T^5 s^{10} + (-25T^5 + 5T^4)s^9 + (231.3050T^5 - 75T^4 + 10T^3)s^8 +$$

$$(-957.2813T^5 + 231.3050T^4 - 50T^3 + 10T^2)s^7 + (1.6568 \times 10^3 T^5 +$$

$$957.2813T^4 - 462.61T^3 + 50T^2 + 5T)s^6 + (-787.6389T^5 - 4.9704 \times 10^3 T^4 +$$

$$1.9146 \times 10^3 T^3 - 462.61T^2 + 75T + 1)s^5 + (3.9382 \times 10^3 T^4 +$$

$$3.3136 \times 10^3 T^3 - 1.9146 \times 10^3 + 231.3050T + 25)s^4 + (-7.8764e \times 10^3 T^3 +$$

$$3.3136 \times 10^3 T^2 - 957.2813T + 231.3050)s^3 + (7.8764 \times 10^3 T^2 - 4.9704 \times 10^3 T +$$

$$957.2813)s^2 + (1.6568 \times 10^3 - 3.9382 \times 10^3 T)s + 787.6389 = 0.$$

$$(3.83)$$

Using Routh's stability criterion, it is concluded that the system will be placed at the margin of the instability with one pair of imaginary roots located at $\omega_{all} = j8.0767$, if $T = T_{all} = 0.1238s$. By utilizing Eq. (3.74), the allowable communication delay h_{all} becomes

$$h_{all} = \frac{2}{\omega_{all}} \tan^{-1}(\omega_{all} T_{all}) = 0.1945s. \tag{3.84}$$

In Figs. 3.22 and 3.23, X and Y components of the position of the follower 8 are shown, where time delays are $h = 0.19s$ and $h = 0.21s$. As it is observed, actual position of the follower 8 ultimately meets its desired position, if h is less than the allowable communication delay ($h = 0.1945s$).

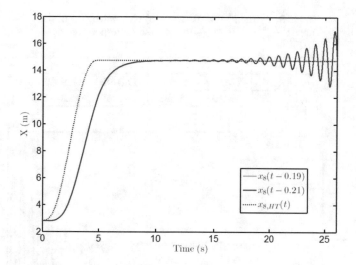

Fig. 3.22 X component of the follower agent 8

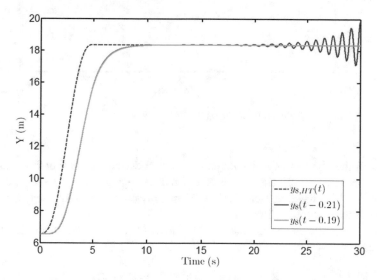

Fig. 3.23 Y component of the follower agent 8

$\beta = 1$: Because all eigenvalues of F are real and $\rho(F) < 1$, eigenvalues of both matrices $I - F$ and $I + F$ are positive and real. Therefore, all roots of the characteristic equation (3.76) are located on the $j\omega$ axis, if

$$T_{all} = \frac{1}{g} = 0.2.$$

Table 3.10 The ith ($i = 1, 2,\ldots, 5$) eigenvalue (η_i) of the matrix $-g^2(I+F)^{-1}(I-F)$ and the corresponding $\omega_i = |\sqrt{\eta_i}|$

	$i = 1$	$i = 2$	$i = 3$	$i = 4$	$i = 5$
η_i	-1.9946	-12.0072	-29.1424	-77.3792	-104.9310
ω_i	1.4123	3.4651	5.3984	8.7965	10.2436

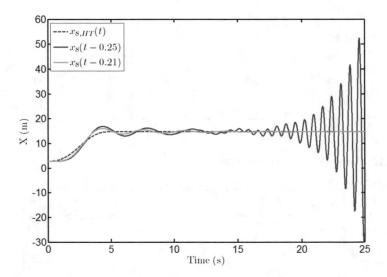

Fig. 3.24 X component of the follower agent 8

It is noted that the matrix F is not symmetric, but the eigenvalues of F are all real. The i^{th}($i = 1, 2,\ldots, 5$) eigenvalue of the matrix $-g^2(I+F)^{-1}(I-F)$ (which is denoted by η_i) is listed in Table 3.10. Additionally crossover frequency $\omega_i = |\sqrt{\eta_i}|$ is given in the last row of Table 3.10.

Given ω_5, allowable communication delay h_{all}, obtained from Eq. (3.74), becomes

$$h_{all} = \frac{2}{\omega_5}\{\tan^{-1}(0.2\omega_5)\} = 0.2180s. \tag{3.85}$$

In Figs. 3.24 and 3.25, X and Y components of the actual position of follower 8 are shown, when $h = 0.21s$ and $h = 0.25s$. As it is observed, $r_8(t)$ asymptotically converges to $r_{8,HT(t)}$ when $h = 0.21s < 0.2180s$. However, evolution of follower 8 is unstable when $h = 0.22s > 0.2180s$.

3.2.3.3 Stability of Delayed MAS Evolution Using Eigen-Analysis

As described above, stability under communication delays in a network of agents can be investigated by the CTCR method. However, the complexity of stability

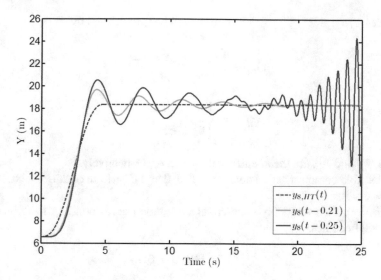

Fig. 3.25 Y component of the follower agent 8

analysis is considerably increased, when the total number of agents is large. To deal with this complexity, an alternative method is presented in this section. The allowable communication delay for each follower agent is formulated based on (i) one of the eigenvalues of the communication matrix putting MAS evolution at the margin of instability, and (ii) the control parameter g_i applied by the follower i. It is noted that these new formulations directly apply the transcendental terms to analyze the stability. This implies that the characteristic equation of the retarded MAS evolution is not approximated by a finite order polynomial.

Suppose that each follower updates the qth component of its current position according to the dynamics (3.18) and (3.19), where $h_{1_i} = h_{2_i} = h_i$. Let

$$z'_{q,i} = \begin{bmatrix} x_{q,n+2}(t-h_i) \\ \vdots \\ x_{q,N}(t-h_i) \end{bmatrix} \in \mathbb{R}^{N-n-1} \tag{3.86}$$

$$u'_{q,i} = \begin{bmatrix} x_{q,1}(t-h_i) \\ \vdots \\ x_{q,n+1}(t-h_i) \end{bmatrix} \in \mathbb{R}^{n+1}, \tag{3.87}$$

then the qth components of the followers' positions are updated by the following first order dynamics:

$$\dot{z}_q = \sum_{j=1}^{N-n-1} [G_j(Az'_{q,j} + Bu'_{q,j}) + \wedge_j(F\dot{z}'_{q,j} + B\ddot{u}'_{q,j})]. \tag{3.88}$$

Matrices $G_j, \wedge_j \in \mathbb{R}^{(N-n-1)\times(N-n-1)}$ in Eq. (3.88) are defined by

$$G_{j_{kl}} = \begin{cases} g_{n+1+j} & k = l = j \\ 0 & else \end{cases}$$

$$\wedge_{j_{kl}} = \begin{cases} \beta & k = l = j \\ 0 & else \end{cases}$$

where $G_{j_{kl}}$ and $\wedge_{j_{kl}}$ are the kl entries of G_j and $\wedge_j$, respectively.

Allowable communication delays for $\beta = 0$ and $\beta = 1$ are obtained next.

$\beta = 0$

If $\beta = 0$, then the delay characteristic equation corresponding to the dynamics (3.88) becomes

$$|sI - GE.A| = 0 \tag{3.89}$$

where

$$GE = diag(g_{n+2}e^{-h_{n+2}s}, \ldots, g_N e^{-h_N s}) \in \mathbb{R}^{(N-n-1)\times(N-n-1)} \tag{3.90}$$

is a positive diagonal matrix. The characteristic Eq. (3.89) can be rewritten in the following form:

$$|\tau(s)I - A| = 0 \tag{3.91}$$

where

$$\tau(s) = sGE^{-1} = diag(\tau_{n+2}(s), \ldots, \tau_N(s)) \in \mathbb{R}^{(N-n-1)\times(N-n-1)} \tag{3.92}$$

and

$$\tau_i(s) = \frac{se^{h_i s}}{g_i}, \tag{3.93}$$

then the following theorem provides an upper bound for the communication delay h_i.

Theorem 3.5. *Suppose that $\beta = 0$ and communication delay h_i in the dynamics (3.88) does not exceed the upper limit*

$$H_i = \frac{H}{g_i} \tag{3.94}$$

where

$$H = min\{\gamma_1, \gamma_2, \ldots, \gamma_{N-n-1}\},$$

$$\gamma_k = \frac{\theta_k - \frac{\pi}{2}}{a_k},$$

a_k *and* θ_k *are the magnitude and argument of* $\lambda_k = a_k e^{j\theta_k}$ *(the* k^{th} *eigenvalue of the matrix A). Then MAS evolution dynamics (3.88) is stable [113].*

Proof. Because $\lambda_k = a_k e^{j\theta_k}$ is the kth eigenvalue of the Hurwitz matrix A,

$$\tau_i(s) = \frac{se^{h_{i,k}s}}{g_i} = a_k e^{j\theta_k} \tag{3.95}$$

satisfies the characteristic Eq. (3.91). Suppose that s crosses the $j\omega$ axis for the first time, when $h_i = h_{i,k}$. This implies that

$$\tau_i(j\omega_{k,i}) = \lambda_k, \tag{3.96}$$

or

$$\frac{j\omega_{k,i}e^{j\omega_{k,i}h_{i,k}}}{g_i} = \frac{\omega_{k,i}e^{j(\omega_{k,i}h_{i,k}+\frac{\pi}{2})}}{g_i} = a_k e^{j\theta_k}. \tag{3.97}$$

Therefore, the crossover frequency $\omega_{k,i}$ and communication delay $h_{i,k}$ are obtained as follows:

$$\omega_{k,i} = g_i a_k \tag{3.98}$$

$$h_{i,k} = \frac{\theta_k - \frac{\pi}{2}}{\omega_{k,i}} = \frac{\gamma_k}{g_i}. \tag{3.99}$$

Let

$$S_i = \{h_{i,k} | \tau_i(j\omega_{k,i}) = \lambda_k, \text{ for } k = 1, 2, \ldots, N-n-1\}$$

be the set of all communication delays satisfying Eq. (3.99) for different eigenvalues λ_k $(k = 1, 2, \ldots, N-n-1)$. Then,

$$H_i = \frac{1}{g_i}min\{\frac{\theta_1 - \frac{\pi}{2}}{a_1}, \ldots, \frac{\theta_{N-n-1} - \frac{\pi}{2}}{a_{N-n-1}}\} \tag{3.100}$$

is considered as the upper bound for the communication delay of the follower i. Thus, it is assured that all roots of the characteristic Eq. (3.89) are at the open left half s-plane and MAS asymptotically converges to a desired final configuration, if the communication delay of the follower i does not exceed the upper bound given in Eq. (3.100).

Corollary. *If the eigenvalues of the matrix A $(a_{N-n-1} < a_{N-n-2} < \cdots < a_1 < 0)$ are all real, then $\theta_k = \pi$ $(k = 1, 2, \ldots, N - n - 1)$ and allowable communication delay of the follower i is determined by Eq. (3.94), where*

$$H = -\frac{\pi}{2a_{N-n-1}}. \tag{3.101}$$

Example 3.5 [113]. Consider again Example 3.4, where $N = 8$, $n = 2$, followers apply the same control gain $g = 5$, and eigenvalues of the matrix A are all real: $\lambda_1 = a_1 = -0.1478$, $\lambda_2 = a_2 = -0.6489$, $\lambda_3 = a_3 = -1.0765$, $\lambda_4 = a_4 = -1.5116$, and $\lambda_5 = a_5 = -1.6152$. Therefore,

$$H = -\frac{\pi}{2a_{N-n-1}} = -\frac{\pi}{2a_5} = \frac{\pi}{2 \times 1.6152} = 0.9752 \tag{3.102}$$

and the allowable communication delay of the follower i becomes

$$h_i = h = \frac{H}{g_i} = \frac{0.9752}{5} = \frac{1}{5}\frac{\pi - \dfrac{\pi}{2}}{1.6152} = 0.1945s, \ i \in V_F. \tag{3.103}$$

As it is observed, the communication delay in Eq. (3.103) is the same as the communication delay obtained from the CTCR method in Eq. (3.84).

$\beta = 1$

When $\beta = 1$, the characteristic equation of MAS evolution dynamics can be rewritten as

$$|sI - (GE \cdot A + sG^{-1}GE \cdot F)| = 0, \tag{3.104}$$

To ensure stability of MAS evolution, all roots of the characteristic Eq. (3.104) are required to be located in the open left half s-plane. It is noticed that the roots of the characteristic Eq. (3.104) are the same as the roots of

$$|\pi(s)I - A| = 0, \tag{3.105}$$

where

$$\pi(s) = diag(\pi_{n+2}(s), \ldots, \pi_N(s)) \in \mathbb{R}^{(N-n-1)\times(N-n-1)} \tag{3.106}$$

and

$$\pi_i(s) = \frac{s(e^{sh_{2i}} - 1)}{s + g_i}, \ i \in V_F. \tag{3.107}$$

In the Theorem 3.6, an upper limit for the communication delay of the follower agent i ($\forall i \in V_F$) is determined, when MAS evolution dynamics is given by Eq. (3.88) and $\beta = 1$.

Theorem 3.6. *MAS evolution dynamics specified by Eq. (3.60) is stable, if h_i (the communication delay of the follower i) does not exceed the upper limit*

$$H_i = min\{h_{i,1}, \ldots, h_{i,N-n-1}\},$$

where

$$h_{i,k} = \frac{1}{\omega_{k,i}} \sin^{-1}\left(\frac{g_i(1-x_k) + y_k\omega_{k,i}}{\omega_{k,i}}\right), \quad k = 1, 2, \ldots, N-n-1, \qquad (3.108)$$

is calculated at $\sigma_k = x_k + y_k j$ (the kth eigenvalue of the matrix $F = A + I$), g_i is the positive control gain applied by the follower i, and the crossover frequency $\omega_{k,i}$ is obtained as

$$\omega_{k,i} = \frac{g_i(y_k + \sqrt{(y_k^2 + (1 - x_k^2 - y_k^2)(y_k^2 + (1 - x_k)^2)})}}{1 - x_k^2 - y_k^2}. \qquad (3.109)$$

Proof. Because σ_k is the kth ($k = 1, 2, \ldots, N-n-1$) eigenvalue of the nonnegative and irreducible matrix $F = A + I$,

$$\pi_i(s) = \frac{s(e^{h_{i,k}s} - 1)}{s + g_i} = -1 + \sigma_k \qquad (3.110)$$

satisfies the characteristic Eq. (3.105). It is noticed that eigenvalues of the matrix F are all located inside the unit desk centered at the origin. Let s in Eq. (3.110) cross $j\omega$ axis for the first time when h_i is increased to $h_{i,k}$. Then Eq. (3.110) can be rewritten as follows:

$$\pi_i(j\omega_{k,i}) + 1 = \frac{g_i + j\omega_{k,i}e^{j\omega_{k,i}h_{i,k}}}{g_i + j\omega_{k,i}} = x_k + y_k j. \qquad (3.111)$$

By equating the real and imaginary parts of both sides of Eq. (3.111), following two relations are obtained:

$$\begin{cases} \omega_{k,i}\sin(\omega_{k,i}h_{i,k}) = g_i(1-x_k) + \omega_{k,i}y_k \\ \omega_{k,i}\cos(\omega_{k,i}h_{i,k}) = \omega_{k,i}x_k + g_iy_k. \end{cases} \qquad (3.112)$$

By solving Eq. (3.112), $\omega_{k,i}$ and $h_{i,k}$ are obtained as given by Eqs. (3.108) and (3.109). Now, let $S_i = \{h_{i,1}, h_{i,2}, \ldots, h_{i,N-n-1}\}$ specify all communication delays satisfying Eq. (3.112) for different eigenvalues of the matrix F. Then the minimum over the set S_i, denoted by H_i, determines the maximum allowable communication delay between a follower i and its in-neighbor agents. Consequently, the MAS evolution governed by the dynamics (3.104) is stable, if the communication delay h_i ($\forall i \in V_F$) is less than the allowable delay H_i.

Corollary. *If the eigenvalues of the matrix F are all real ($y_k = 0, k = 1,\ 2,\ldots$, $N - n - 1$), then $\omega_{k,i}$ and $h_{i,k}$ simplify to:*

$$\omega_{k,i} = g_i \sqrt{\frac{1 - x_k}{1 + x_k}} \tag{3.113}$$

$$h_{i,k} = \frac{1}{g_i} \sqrt{\frac{1 + x_k}{1 - x_k}} \cos^{-1} x_k. \tag{3.114}$$

Note that $h_{i,k}$ is nonnegative and increasing for $x_k \in (-1,1)$. Therefore, the smallest eigenvalue of the matrix F specifies the allowable delay H_i for the follower $i \in V_F$.

Example 3.6. Consider Example 3.4, where the eigenvalues of the matrix F are all real: $\sigma_1 = x_1 = 0.8522$, $\sigma_2 = x_2 = 0.3511$, $\sigma_3 = x_3 = -0.0765$, $\sigma_4 = x_4 = -0.5116$, and $\sigma_5 = x_5 = -0.6152$. Therefore,

$$H_i = \frac{1}{g} \sqrt{\frac{1 + x_5}{1 - x_5}} \cos^{-1}(x_5) = \frac{1}{5} \sqrt{\frac{1 - 0.6152}{1 + 0.6152}} \cos^{-1}(-0.6152) = 0.2181s. \tag{3.115}$$

specifies the allowable communication delay for the follower i. As it is seen, H_i obtained from Eqs. (3.115) and (3.85) are equal.

3.2.4 MAS Evolution Dynamics-First Order Kinematic Model-Method 2

Suppose $n + 1$ leaders at the vertices of the leading polytope in $\mathbb{R}^n$ evolve independently, where their positions satisfy the rank condition (2.3) at any time t. Let $r_i \in \mathbb{R}^n$ (current position of the follower $i \in V_F$) be expanded as the linear combination of $r_{i1}, r_{i2}, \ldots, r_{in+1}$ (current positions of in-neighbor agents $i_1, i_2, \ldots, i_{n+1}$) as follows:

$$r_i = \sum_{k=1}^{n+1} \varpi_{i,k}(t) r_{ik}. \tag{3.116}$$

It is assumed that positions of the adjacent agents $i_1, i_2, \ldots, i_{n+1}$ satisfy the following rank condition:

$$\forall t \geq t_0, Rank \left[r_{i2} - r_{i1} \ \ldots \ r_{i2} - r_{in+1} \right] = n. \tag{3.117}$$

Then the time varying weight ϖ_{i,i_k} is unique, and

$$\sum_{k=1}^{n+1} \varpi_{i,i_k}(t) = 1. \tag{3.118}$$

By considering Eqs. (3.116), (3.117), and (3.118) the transient weight ϖ_{i,i_k} is obtained by solving the following set of n linear algebraic equations:

$$r_i = P_i W_i + r_{i_{n+1}} \tag{3.119}$$

where $W_i(t) = [\varpi_{i,i_1} \ \ldots \ \varpi_{i,i_n}]^T \in \mathbb{R}^n$, and

$$P_i = \begin{bmatrix} x_{1,i_1} - x_{1,i_{n+1}} & x_{1,i_2} - x_{1,i_{n+1}} & \cdots & x_{1,i_n} - x_{1,i_{n+1}} \\ x_{2,i_1} - x_{2,i_{n+1}} & x_{2,i_2} - x_{2,i_{n+1}} & \cdots & x_{2,i_n} - x_{2,i_{n+1}} \\ \vdots & \vdots & \ddots & \vdots \\ x_{n,i_1} - x_{n,i_{n+1}} & x_{n,i_2} - x_{n,i_{n+1}} & \cdots & x_{n,i_n} - x_{n,i_{n+1}} \end{bmatrix}. \tag{3.120}$$

Since the transient weight ϖ_{i,i_k} ($i \in V_F$, $k = 1, 2, \ldots, n+1$) can be uniquely determined at any time t, homogeneous deformation of the MAS is achieved if the follower $i \in V_F$ updates its position such that ϖ_{i,i_k} is as close as possible to the communication weight w_{i,i_k} which is obtained based on agents' initial positions by applying Eq. (3.6).

Theorem 3.7. *Consider an MAS consisting of N agents, where (i) agents $1, 2, \ldots, n+1$ are the leaders moving independently and leaders' positions satisfy the rank condition (2.3), (ii) each follower i is initially placed inside a communication polytope whose vertices are occupied by the in-neighbor agents $i_1, i_2, \ldots, i_{n+1}$, and (iii) positions of the in-neighbor agents satisfy the rank condition (3.117). If the weight vector $W_i(t) = [\varpi_{i,i_1} \ \ldots \ \varpi_{i,i_n}]^T \in \mathbb{R}^n$ is updated by*

$$\dot{W}_i = -g_i(t)W_i + g_i(t)W_{i0} \tag{3.121}$$

where $W_{i0} = W_i(0) = [w_{i,i_1} \ \ldots \ w_{i,i_n}]^T \in \mathbb{R}^{n+1}$, and $g_i(t)$ satisfies the inequality

$$\forall t \geq 0, \ g_i(t) > max(|\lambda_k[(\dot{P}_i P_i^{-1})^T + \dot{P}_i P_i^{-1}]|), \tag{3.122}$$

then the initial formation of the MAS asymptotically converges to a final formation given by the homogeneous mapping, when the leaders stop. Note that λ_k is the kth eigenvalue of the matrix $[\dot{P}_i P_i^{-1}]$.

Proof. Taking the time derivative from Eq. (3.119) results in

$$\dot{r}_i = \dot{P}_i W_i + P_i \dot{W}_i + \dot{r}_{i_{n+1}} \tag{3.123}$$

Now, replacing $\dot{W}_i$ in Eq. (3.123) by Eq. (3.121) leads to

$$\dot{r}_i = (\dot{P}_i - g_i(t)P_i)W_i + g_i(t)P_i W_{i0} + \dot{r}_{i_{n+1}}. \tag{3.124}$$

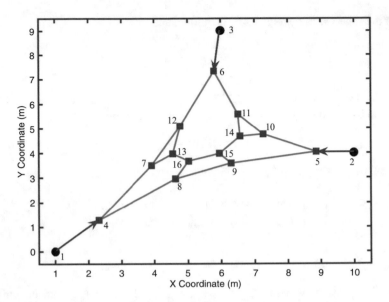

Fig. 3.26 Agents' initial positions and the communication graph in Example 3.7

By letting $W_i = P_i^{-1}(r_i - r_{i_{n+1}})$, Eq. (3.124) is converted to

$$\dot{r}_i = (\dot{P}_i P_i^{-1} - g_i(t)I)r_i + \dot{r}_{i_{n+1}} + (\dot{P}_i P_i^{-1} - g_i(t)P_i)r_{i_{n+1}} + g_i(t)P_i W_{i0}. \qquad (3.125)$$

Thus, MAS evolution is stable if

$$g_i(t) > max(|\lambda_k[(\dot{P}_i P_i^{-1})^T + \dot{P}_i P_i^{-1}]|), \quad \forall t \geq 0.$$

It is noticed that the equilibrium state of the dynamics (3.125), given by

$$r_{is} = P_i W_{i0} + r_{i_{n+1}}, \qquad (3.126)$$

is obtained if $\dot{r}_i$, $\dot{P}_i$, $\dot{r}_{i_{n+1}}$ vanish. It is observed that $W_i(t)$ converges to $W_{i_0}(t)$ when leaders stop. Therefore, final value of $\varpi_{i,j}(t)(i \in V_F, j \in N_i)$ converges to $w_{i,j}$ and MAS final formation is a homogeneous transformation of the initial distribution of the agents.

Example 3.7. Consider an MAS containing 16 agents with the initial distribution shown in Fig. 3.26. As it is seen, agents 1, 2, and 3 are the leaders, and the agents 4, 5, ..., 16 are the followers. Followers apply the graph shown in Fig. 3.26 to acquire the desired position given by a homogeneous deformation through local communication. Communication weights as well as agents' initial positions are listed in Table 3.11.

Table 3.11 Initial positions and communication weights associated with the formation shown in Fig. 3.26

i	X_i	Y_i	i_1	i_2	i_3	w_{i,i_1}	w_{i,i_2}	w_{i,i_3}
1	1.00	0.00	–	–	–	–	–	–
2	10.00	4.00	–	–	–	–	–	–
3	6.00	9.00	–	–	–	–	–	–
4	2.33	1.28	1	7	8	0.60	0.18	0.22
5	8.86	4.04	2	9	10	0.65	0.19	0.16
6	5.80	7.34	3	11	12	0.55	0.20	0.25
7	3.92	3.50	4	12	13	0.32	0.35	0.33
8	4.64	2.95	4	9	16	0.29	0.31	0.40
9	6.30	3.58	5	8	15	0.30	0.40	0.30
10	7.26	4.76	5	11	14	0.31	0.31	0.38
11	6.51	5.57	6	10	14	0.33	0.28	0.39
12	4.78	5.09	6	7	13	0.39	0.41	0.20
13	4.56	3.96	7	12	16	0.37	0.25	0.38
14	6.57	4.67	10	11	15	0.36	0.26	0.38
15	5.96	3.97	9	14	16	0.33	0.33	0.34
16	5.03	3.67	8	13	15	0.29	0.39	0.32

The desired weight vector W_{i0}, the vector $r_{in+1}(t)$, and the matrix $P_i(t)$ (that are consistent with the communication graph and agents' positions shown in Fig. 3.26) are obtained as listed in Table 3.12.

Paths of the leaders 1, 2, and 3 are shown in Fig. 3.27, where all leaders finally stop in 20s. The follower agent $i \in V_F$ chooses

$$g_i(t) = 5 + max(|\lambda_k[(\dot{P}_i P_i^{-1})^T + \dot{P}_i P_i^{-1}]|).$$

Variation of g_{15} versus time t is shown in Fig. 3.28 (g_{15} is the control gain applied by the follower $i = 15$.). Also, the first (X) and second (Y) components of actual position of the follower 15 as a function of time t is shown by continuous curves in Fig. 3.29. Note that dotted curves in Fig. 3.29 illustrate the first (X) and second (Y) components of the desired position of the follower 15 given by a homogeneous transformation.

In Fig. 3.30 transient weights $\varpi_{15,9}(t)$, $\varpi_{15,14}(t)$, and $\varpi_{15,16}(t)$ are shown. It is noted that $\varpi_{15,9}(t)$, $\varpi_{15,14}(t)$, and $\varpi_{15,16}(t)$ are calculated by using Eq. (3.119) based on the actual positions of the follower 15 and the in-neighbor agents 9, 14, and 16.

As it is seen in Fig. 3.30, initial and final values of the transient weights $\varpi_{15,9}(t)$, $\varpi_{15,14}(t)$, and $\varpi_{15,16}(t)$ are the same as the communication weights $w_{15,9} = 0.33$, $w_{15,14} = 0.33$, and $w_{15,16} = 0.34$ (See Table 3.13.) It is true for all other follower agents as well which implies that the final formation is a homogeneous transformation of the initial configuration of the MAS. In Fig. 3.31, configurations of the agents at three different sample times $t = 5s$, $t = 13s$, and $t = 25s$ are shown by $\square$, $\bigcirc$, and $\triangle$, respectively.

Table 3.12 W_{i0}, $r_{in+1}(t)$, and $P_i(t)$ of the follower i that are consistent with agents' formation shown in Fig. 3.26

i	W_{i0}			$r_{in+1}(t)$		$P_i(t)$	
4	$W_{40} =$	$\begin{matrix} w_{4,1} \\ w_{4,7} \end{matrix} =$	$\begin{matrix} 0.60 \\ 0.18 \end{matrix}$	$r_{i_3} = r_8 =$	$\begin{matrix} x_8 \\ y_8 \end{matrix}$	$P_4 =$	$\begin{matrix} x_1 - x_8 & x_7 - x_8 \\ y_1 - y_8 & y_7 - y_8 \end{matrix}$
5	$W_{50} =$	$\begin{matrix} w_{5,2} \\ w_{5,9} \end{matrix} =$	$\begin{matrix} 0.65 \\ 0.19 \end{matrix}$	$r_{i_3} = r_{10} =$	$\begin{matrix} x_{10} \\ y_{10} \end{matrix}$	$P_5 =$	$\begin{matrix} x_2 - x_{10} & x_9 - x_{10} \\ y_2 - y_{10} & y_9 - y_{10} \end{matrix}$
6	$W_{60} =$	$\begin{matrix} w_{6,3} \\ w_{6,11} \end{matrix} =$	$\begin{matrix} 0.55 \\ 0.20 \end{matrix}$	$r_{i3} = r_{12} =$	$\begin{matrix} x_{12} \\ y_{12} \end{matrix}$	$P_6 =$	$\begin{matrix} x_3 - x_{12} & x_{11} - x_{12} \\ y_3 - y_{12} & y_{11} - y_{12} \end{matrix}$
7	$W_{70} =$	$\begin{matrix} w_{7,4} \\ w_{7,12} \end{matrix} =$	$\begin{matrix} 0.32 \\ 0.35 \end{matrix}$	$r_{i3} = r_{13} =$	$\begin{matrix} x_{13} \\ y_{13} \end{matrix}$	$P_7 =$	$\begin{matrix} x_4 - x_{13} & x_{12} - x_{13} \\ y_4 - y_{13} & y_{12} - y_{13} \end{matrix}$
8	$W_{80} =$	$\begin{matrix} w_{8,4} \\ w_{8,9} \end{matrix} =$	$\begin{matrix} 0.29 \\ 0.31 \end{matrix}$	$r_{i3} = r_{16} =$	$\begin{matrix} x_{16} \\ y_{16} \end{matrix}$	$P_8 =$	$\begin{matrix} x_4 - x_{16} & x_9 - x_{16} \\ y_4 - y_{16} & y_9 - y_{16} \end{matrix}$
9	$W_{90} =$	$\begin{matrix} w_{9,5} \\ w_{9,8} \end{matrix} =$	$\begin{matrix} 0.30 \\ 0.40 \end{matrix}$	$r_{i3} = r_{15} =$	$\begin{matrix} x_{15} \\ y_{15} \end{matrix}$	$P_9 =$	$\begin{matrix} x_5 - x_{15} & x_8 - x_{15} \\ y_5 - y_{15} & y_8 - y_{15} \end{matrix}$
10	$W_{100} =$	$\begin{matrix} w_{10,5} \\ w_{10,11} \end{matrix} =$	$\begin{matrix} 0.31 \\ 0.31 \end{matrix}$	$r_{i3} = r_{14} =$	$\begin{matrix} x_{14} \\ y_{14} \end{matrix}$	$P_{10} =$	$\begin{matrix} x_5 - x_{14} & x_{11} - x_{14} \\ y_5 - y_{14} & y_{11} - y_{14} \end{matrix}$
11	$W_{110} =$	$\begin{matrix} w_{11,6} \\ w_{11,10} \end{matrix} =$	$\begin{matrix} 0.33 \\ 0.28 \end{matrix}$	$r_{i3} = r_{14} =$	$\begin{matrix} x_{14} \\ y_{14} \end{matrix}$	$P_{11} =$	$\begin{matrix} x_6 - x_{14} & x_{10} - x_{14} \\ y_6 - y_{14} & y_{10} - y_{14} \end{matrix}$
12	$W_{120} =$	$\begin{matrix} w_{12,6} \\ w_{12,7} \end{matrix} =$	$\begin{matrix} 0.39 \\ 0.41 \end{matrix}$	$r_{i3} = r_{13} =$	$\begin{matrix} x_{13} \\ y_{13} \end{matrix}$	$P_{12} =$	$\begin{matrix} x_6 - x_{13} & x_7 - x_{13} \\ y_6 - y_{13} & y_7 - y_{13} \end{matrix}$
13	$W_{130} =$	$\begin{matrix} w_{13,7} \\ w_{13,12} \end{matrix} =$	$\begin{matrix} 0.37 \\ 0.25 \end{matrix}$	$r_{i3} = r_{16} =$	$\begin{matrix} x_{16} \\ y_{16} \end{matrix}$	$P_{13} =$	$\begin{matrix} x_7 - x_{16} & x_{12} - x_{16} \\ y_7 - y_{16} & y_{12} - y_{16} \end{matrix}$
14	$W_{140} =$	$\begin{matrix} w_{14,10} \\ w_{14,11} \end{matrix} =$	$\begin{matrix} 0.36 \\ 0.26 \end{matrix}$	$r_{i3} = r_{15} =$	$\begin{matrix} x_{15} \\ y_{15} \end{matrix}$	$P_{14} =$	$\begin{matrix} x_{10} - x_{15} & x_{11} - x_{15} \\ y_{10} - y_{15} & y_{11} - y_{15} \end{matrix}$
15	$W_{150} =$	$\begin{matrix} w_{15,9} \\ w_{15,14} \end{matrix} =$	$\begin{matrix} 0.33 \\ 0.33 \end{matrix}$	$r_{i3} = r_{16} =$	$\begin{matrix} x_{16} \\ y_{16} \end{matrix}$	$P_{15} =$	$\begin{matrix} x_9 - x_{16} & x_{14} - x_{16} \\ y_9 - y_{16} & y_{14} - y_{16} \end{matrix}$
16	$W_{160} =$	$\begin{matrix} w_{16,8} \\ w_{16,13} \end{matrix} =$	$\begin{matrix} 0.29 \\ 0.39 \end{matrix}$	$r_{i3} = r_{15} =$	$\begin{matrix} x_{15} \\ y_{15} \end{matrix}$	$P_{16} =$	$\begin{matrix} x_8 - x_{15} & x_{13} - x_{15} \\ y_8 - y_{15} & y_{13} - y_{15} \end{matrix}$

3.3 Homogeneous Deformation of an MAS Using Preservation of Volumetric Ratios

The above strategy relying on minimum interagent communication limits the number of agents and the interagent communication. Also, the magnitude of the real part of the smallest eigenvalue of the matrix A may not be high enough, when interagent interaction is minimum. Therefore, deviation from the desired position, given by a homogeneous deformation, may be considerable during MAS evolution, although achieving the final homogeneous configuration is assured. On the other hand, if we raise the number of interagent communications, then communication weights cannot be assigned uniquely based on the agents' initial positions, unless predetermined communication weights are selected for the followers. So, it is

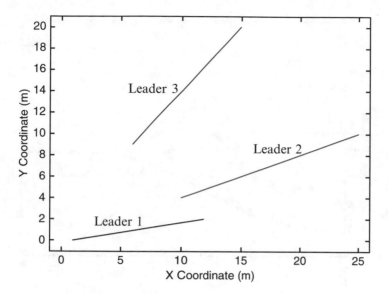

Fig. 3.27 Paths of the leaders in Example 3.7

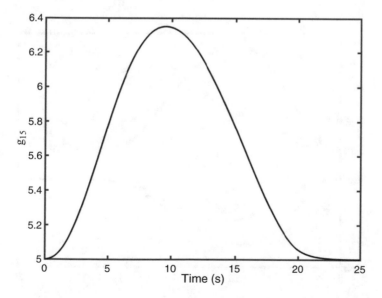

Fig. 3.28 Gain $g_{15}(t)$ applied by the follower 15 in Example 3.7

interesting to know how a homogeneous transformation can be acquired by an MAS, when the follower i is allowed to interact with $p_i > n + 1$ local agents to evolve. In this section we investigate this problem and show how homogeneous transformation of an MAS can be achieved by preserving some volumetric ratios, where followers are not restricted to interact only with $n + 1$ neighboring agents.

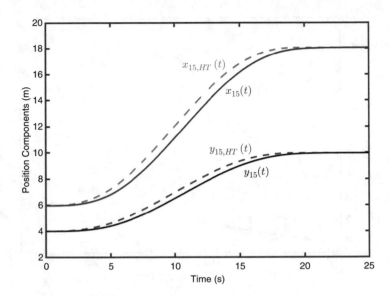

Fig. 3.29 X and Y components of the follower 15

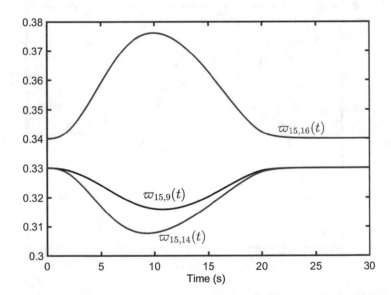

Fig. 3.30 Time varying weights $\varpi_{15,9}(t)$, $\varpi_{15,14}(t)$, and $\varpi_{15,16}(t)$

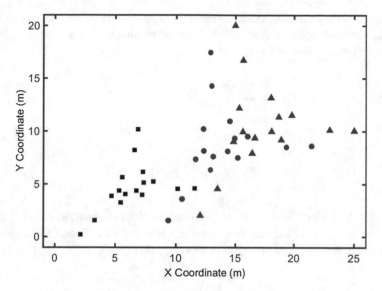

Fig. 3.31 MAS Formations at $t = 5s$, $t = 13s$, and $t = 25s$

Volumetric Weight Ratios Consider a communication polytope $\Omega_i \subset \mathbb{R}^n$, that is the union of m_i different sub-polyhedra $\Omega_{i,1}$, $\Omega_{i,2}$, ..., Ω_{i,m_i}, where there is no intersection between any two sub-polyhedra inside Ω_i. Notice that vertices of the polytope Ω_i is occupied by $p_i \geq n+1$ in-neighbor agents of the follower $i \in V_F$. Let

$$aw_{ij} = \frac{v_{i,j}}{v_i},\qquad (3.127)$$

be the ratio of the volume of the sub-polyhedron $\Omega_{i,j}$ ($j = i_1, i_2, \ldots, i_{m_i}$) to the net volume of the polyhedron Ω_i. Volume of the sub-polyhedron $\Omega_{i,j}$ is computed by

$$v_{i,j} = \frac{1}{n!}\begin{vmatrix} x_{1,i} & x_{2,i} & \cdots & x_{n,i} & 1 \\ x_{1,i_1} & x_{2,i_1} & \cdots & x_{n,i_1} & 1 \\ x_{1,i_2} & x_{2,i_2} & \cdots & x_{n,i_2} & 1 \\ \vdots & \vdots & \ddots & \vdots & \vdots \\ x_{1,i_{n+1}} & x_{2,i_{n+1}} & \cdots & x_{n,i_{n+1}} & 1 \end{vmatrix} = \sum_{q=1}^{n} O_{q,i,j}x_{q,i} + P_{ij}(t) \qquad (3.128)$$

and

$$v_i = \sum_{j=1}^{m_i} v_{i,j}. \qquad (3.129)$$

It is noted that $v_{i,j}$ must be positive. This can easily be assured by interchanging two rows of the determinant if needed. Since the net volume $v_i(t)$ at any time $t \geq t_0$ does not depend on position of the agent i,

$$\sum_{j=1}^{m_i} \sum_{q=1}^{n} O_{q,i,j} = 0, \ q = 1, 2, \ldots, n \qquad (3.130)$$

and

$$v_i(t) = \sum_{j=1}^{m_i} P_{ij}(t). \qquad (3.131)$$

Remark 3.7. Under a homogeneous deformation, the volume ratio aw_{ij} ($i \in V_F$, $j \in N_i = \{i_1, i_2, \ldots, i_{p_i}\}$) remains time invariant at any time t. Therefore, aw_{ij} can be calculated based on initial configuration of the polytope Ω_i,

$$AW_{ij} = \frac{V_{i,j}}{V_i}, \qquad (3.132)$$

where $V_{i,j} = v_{i,j}(t_0)$ and $V_i = v_i(t_0)$ are initial volume of the jth ($j = i_1, i_2, \ldots, i_{m_i}$) sub-polyhedron and the polytope Ω_i, respectively. The ratio AW_{ij} is called *volumetric weight ratio*.

2-D Deformation (Area Weights): Schematic of a communication polygon Ω_i is shown in Fig. 3.32. As seen in Fig. 3.32, vertices of the polygon Ω_i are occupied by the in-neighbor agents $i_1, i_2, \ldots, i_{m_i}$. Also, the polygon Ω_i is the union of m_i triangles, with initial areas $A_{i,1}, A_{i,2}, \ldots, A_{i,m_i}$, where

$$A_i = A_{i,1} + A_{i,2} + \cdots + A_{i,m_i}$$

denotes the net area of the polygon.

Fig. 3.32 Schematic of the polygon Ω_i that consists of m_i sub-triangles

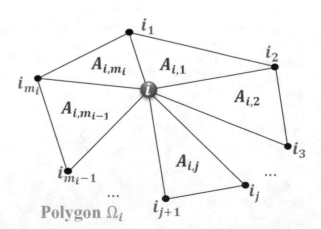

For homogeneous deformation in a plane, the ratio

$$AW_{ij} = \frac{A_{i,j}}{A_i} \qquad (3.133)$$

is the jth area weight of the follower i. Note that AW_{ij} remains time-invariant, if the polygon Ω_i deforms under a homogeneous mapping. Note that

$$\sum_{j=1}^{m_i} AW_{i,j} = 1 \qquad (3.134)$$

and

$$A_{i,j} = \frac{1}{2} \begin{vmatrix} X_i & Y_i & 1 \\ X_{i_{j+1}} & Y_{i_{j+1}} & 1 \\ X_{i_j} & Y_{i_j} & 1 \end{vmatrix}, \qquad (3.135)$$

where (X_i, Y_i), $(X_{i_{j+1}}, Y_{i_{j+1}})$, and (X_{i_j}, Y_{i_j}) are initial positions of agents i, i_{j+1}, and i_j, respectively.

Communication Protocol: Consider an MAS that consists of N agents and moves collectively in $\mathbb{R}^n$. Agents 1, 2, $\ldots$, $N_l (l \geq n+1)$ are the leaders located at the vertices of the leading polytope, where they are transformed under a homogeneous mapping. The remaining agents are the followers updating their positions through communication with some neighboring agents. It is noted that every follower is allowed to interact with $p_i \geq n+1$ local agents, where positions of the in-neighbor agents satisfy the following rank condition:

$$\forall t \geq t_0, Rank \left[r_{i_2} - r_{i_1} \ \ldots \ r_{i_{p_i}} - r_{i_1} \right] = n. \qquad (3.136)$$

Shown in Fig. 3.33 is a typical communication graph applied by an MAS to move collectively in a plane. In Fig. 3.33 rounded nodes represent leaders, where they are located at the vertices of the leading polygon and numbered by 1, 2, $\ldots$, 5. Furthermore, followers shown by squares are all placed inside the leading polygon and numbered by 6, 5, $\ldots$, 14. Leader-follower interaction is shown by an arrow terminated to the follower. This is because leaders move independently, but position of a leader is tracked by a follower. Furthermore, follower-follower communication is shown by a nondirected edge. This implies bidirectional communication between two in-neighbor followers.

To satisfy the rank condition (3.136), the follower agent $i \in V_F$ communicates with at least 3 in-neighbor agents, where in-neighbor agents are not all aligned in the initial configuration. Also, every follower agent i is initially placed inside the ith ($i = 1, 2, \ldots, N$) communication polygon. In Table 3.13, agents' initial positions

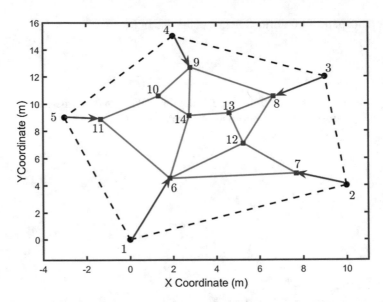

Fig. 3.33 A typical communication graph applied for planar continuum of an MAS using the method of preservation of area ratios

and the corresponding area weights, that are consistent with the configuration and the graph shown in Fig. 3.33, are listed.

***State of Homogeneous Transformation*:** Suppose that leader agents of the MAS are transformed as homogeneous deformation, the initial and current positions of the leader i ($i = 1, 2, \ldots, N_l$) satisfy Eq. (2.2). It is aimed that followers acquire the desired positions defined by a homogeneous transformation by local communication through preservation of volumetric ratios. For this purpose, the cost

$$J_i(t) = \frac{1}{2} \sum_{j=1}^{m_i} (v_{i,j}(t) - AW_{i,j} v_i(t))^2, \tag{3.137}$$

is considered for evolution of the follower $i \in V_F$. The cost function $J_i(t)$ penalizes deviation of the follower i from the desired position defined by a homogeneous transformation.

Remark 3.8. If agents (followers and leaders) are transformed as homogeneous deformation, the cost functions $J_i(t)$ ($\forall i \in V_F$) vanish simultaneously at any $t \in [t_0, T]$. Notice that T is the time when leaders stop.

Followers can determine the desired position through local communication with p_i in-neighbor agents, where the follower agent $i \in V_F$ updates its current position such that $J_i(t)$ is minimized at any time t. The local minimum of the cost function $J_i(t)$ is obtained at

Table 3.13 Followers' initial positions and initial area weights associated with the MAS configuration shown in Fig. 3.33

i	X_i	Y_i	i_1	i_2	i_3	i_4	i_5	AW_{i,i_1}	AW_{i,i_2}	AW_{i,i_3}	AW_{i,i_4}	AW_{i,i_5}
1	0.0000	0.0000	–	–	–	–	–	–	–	–	–	–
2	10.0000	4.0000	–	–	–	–	–	–	–	–	–	–
3	9.0000	12.0000	–	–	–	–	–	–	–	–	–	–
4	2.0000	15.0000	–	–	–	–	–	–	–	–	–	–
5	−3.0000	9.0000	–	–	–	–	–	–	–	–	–	–
6	1.8474	4.5152	1	7	12	14	11	0.2745	0.1469	0.1424	0.1978	0.2384
7	7.6986	4.8704	2	12	6	–	–	0.1304	0.6087	0.2609	–	–
8	6.6566	10.5475	3	9	13	12	–	0.3386	0.2933	0.1727	0.1954	–
9	2.8193	12.6743	8	4	10	14	–	0.2291	0.1663	0.1651	0.4395	–
10	1.3242	10.5533	9	11	14	–	–	0.2143	0.4286	0.3571	–	–
11	−1.3275	8.8683	5	6	10	–	–	0.2549	0.6275	0.1176	–	–
12	5.2544	7.0789	7	8	13	6	–	0.2894	0.1346	0.2313	0.3448	–
13	4.6015	9.3031	8	14	12	–	–	0.1667	0.3667	0.4667	–	–
14	2.7536	9.1197	13	9	10	6	–	0.2345	0.1851	0.2819	0.2984	–

$$r_{i,d}(t) = \sum_{q=1}^{n} x_{q,i,d}(t)\hat{\mathbf{e}}_q, \tag{3.138}$$

where $x_{q,i,d}(t)$ is obtained from

$$
\begin{bmatrix} x_{1,i,d} \\ x_{2,i,d} \\ \vdots \\ x_{n,i,d} \end{bmatrix} = -
\begin{bmatrix}
\sum_{j=1}^{m_i} O_{1,i,j}^2 & \sum_{j=1}^{m_i} O_{1,i,j}O_{2,i,j} & \cdots & \sum_{j=1}^{m_i} O_{1,i,j}O_{n,i,j} \\
\sum_{j=1}^{m_i} O_{2,i,j}O_{1,i,j} & \sum_{j=1}^{m_i} O_{2,i,j}^2 & \cdots & \sum_{j=1}^{m_i} O_{2,i,j}O_{n,i,j} \\
\vdots & \vdots & \ddots & \vdots \\
\sum_{j=1}^{m_i} O_{n,i,j}O_{1,i,j} & \sum_{j=1}^{m_i} O_{n,i,j}O_{2,i,j} & \cdots & \sum_{j=1}^{m_i} O_{n,i,j}^2
\end{bmatrix}^{-1}
$$
$$
\begin{bmatrix}
\sum_{j=1}^{m_i} (O_{1,i,j}P_{ij} - v_i AW_{i,j}O_{1,i,j}) \\
\sum_{j=1}^{m_i} (O_{2,i,j}P_{ij} - v_i AW_{i,j}O_{2,i,j}) \\
\vdots \\
\sum_{j=1}^{m_i} (O_{n,i,j}P_{ij} - v_i AW_{i,j}O_{n,i,j})
\end{bmatrix}. \tag{3.139}
$$

Notice that

$$\frac{\partial J_i}{\partial x_{q,i}} = 0$$

is satisfied at $r_i = r_{i,d} = \sum_{q=1}^{n} x_{q,i,d}(t)\hat{\mathbf{e}}_q$.

Followers' Dynamics: It is assumed that the follower $i \in V_F$ updates its current position according to the following first order dynamics:

Fig. 3.34 Level curves of the local cost function $J_i(t)$

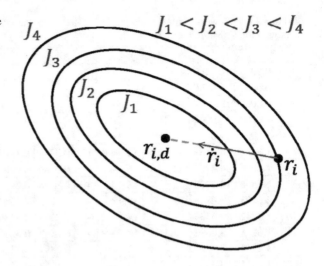

$$\dot{r}_i = g_i(r_{i,d} - r_i), \ \forall i \in V_F. \tag{3.140}$$

Because $g_i \in \mathbb{R}_+$ is constant, set of equilibrium points, denoted by $r_i = r_{i,d}$ ($\forall i \in V_F$), are locally stable. As shown in Fig. 3.34, the level surfaces $J_i = constant$ represent ellipsoids all centered at $r_{i,d}$, thus, they are convex and if the position of the follower $i \in V_F$ is updated according to Eq. (3.140), then $\dot{J}_i < 0$ in the vicinity of $r_{i,d}$. Hence, J_i remains bounded and it asymptotically converges to zero when the leaders stop. This implies that final formation is a homogeneous deformation of the initial configuration.

Example 3.8. Consider the agents' initial configuration shown in Fig. 3.33. Leaders 1, 2,..., 5 are located at the vertices of the leading pentagon, while followers are all distributed inside the leading pentagon. Note that leaders move on the paths shown in Fig. 3.35, where they are initially at $(X_1, Y_1) = (0,0)$, $(X_2, Y_2) = (10,4)$, $(X_3, Y_3) = (9,12)$, $(X_4, Y_4) = (2,15)$, and $(X_5, Y_5) = (-3,9)$ and eventually stop at $(X_{1,F}, Y_{1,F}) = (45,35)$, $(X_{2,F}, Y_{2,F}) = (60,49)$, $(X_{3,F}, Y_{3,F}) = (69,42)$, $(X_{4,F}, Y_{4,F}) = (65.75, 28.33)$, and $(X_{5,F}, Y_{5,F}) = (53.25, 24)$ at time $t = 20s$. Notice that leaders deform under a homogeneous transformation, where entries of the Jacobian Q (Q_{11}, Q_{12}, Q_{21}, and Q_{22}) and the vector D (D_1 and D_2) are uniquely related to the X and Y components of the positions of the leaders 1, 2, and 3 according to Eq. (2.21).

   Initial positions of the followers and the corresponding area weights are all listed in Table 3.13. It is noted that the follower $i \in V_F$ updates its position according to Eq. (3.140), where $g_i = g = 10$ ($\forall i \in V_F$). In Fig. 3.36, X and Y components of the actual position of the agents 14 are shown by continuous curves. Additionally, X and Y components of the desired position of the follower 14, given by the homogeneous transformation, are illustrated by the dotted curves.

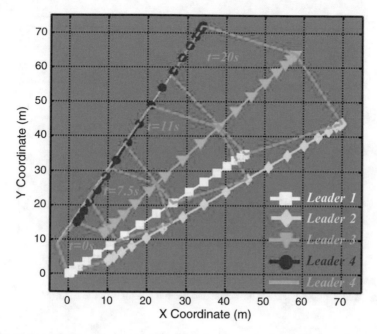

Fig. 3.35 Paths of the leader agents in Example 3.8; Configurations of the leading polygon at different sample time $t = 0s$, $t = 7s$, $t = 11s$, and $t = 20s$

Configurations of the MAS at different times $t = 0s$, $t = 7s$, $t = 11s$, $t = 18s$, and $t = 25s$ are shown in Fig. 3.37.

In Fig. 3.38, the transient area weights of the follower 14, that are calculated by using Eq. (3.127), are shown versus time. As it is seen, the initial and final area weights are identical. Therefore, follower 14 ultimately reaches the desired position defined by the homogeneous deformation in Example 3.8.

3.4 Comparison of the Different Continuous-Time Algorithms for Homogeneous Deformation of an MAS

In this section, evolution of an MAS consisting 14 agents (3 leaders and 11 followers) is investigated, where followers apply different algorithms discussed in Chapters 2 and 3. The main objective is to evaluate the convergence rate of the followers' evolution under different proposed methods in Chapter 3.

Example 3.9. Consider an MAS consisting of 3 leaders and 11 followers with the initial positions shown in Fig. 3.39 as listed in Table 3.14. The parameter $\alpha_{i,k}$ ($\forall i \in V_F$ and $k = 1, 2, 3$) is also obtained by using Eq. (2.18) as listed in the last three columns of Table 3.14.

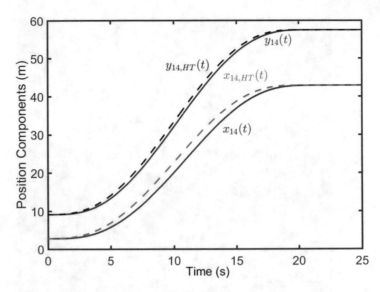

Fig. 3.36 X and Y coordinates of the actual and desired positions of the follower 14

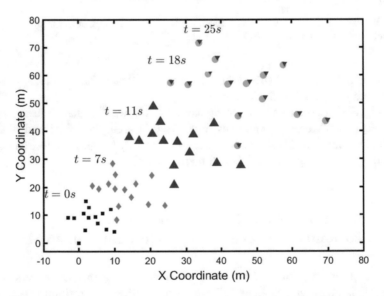

Fig. 3.37 Formations of the MAS at sample times $t = 0s$, $t = 7s$, $t = 11s$, $t = 18s$, and $t = 25s$

Leaders are initially placed at $(X_1, Y_1) = (10, 10)$, $(X_2, Y_2) = (15, 10)$, and $(X_3, Y_3) = (15, 20)$ at $t = 0s$. Let leaders choose the paths shown in Fig. 4.13, and come to rest at $(X_{1,F}, Y_{1,F}) = (30, 20)$, $(X_{2,F}, Y_{2,F}) = (45, 15)$, and $(X_{3,F}, Y_{3,F}) = (35, 25)$ at $t = 20s$.

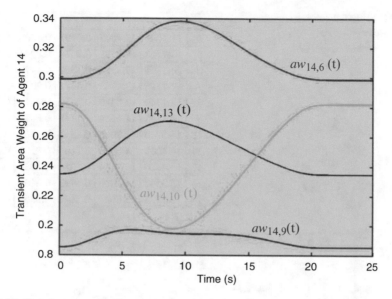

Fig. 3.38 Transient area weights $aw_{14,6}$, $aw_{14,9}$, $aw_{14,10}$, and $aw_{14,13}$

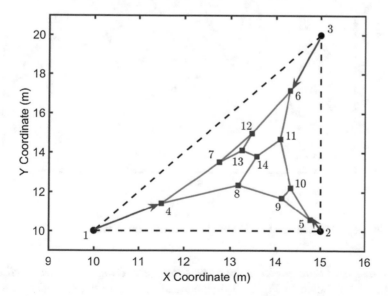

Fig. 3.39 The graph defining interagent communication in Example 3.9

Followers' Evolution under No Interagent Communication: In this scenario, followers acquire homogeneous deformation using no inter-agent communication approach discussed in Chapter 2. Each follower $i \in V_F$ can reach the desired position $r_{i,HT}$ only by knowing positions of the leaders at $t \in [0, 20]s$ and parameters $\alpha_{i,1}$, $\alpha_{i,2}$, and $\alpha_{i,3}$. Given parameters $\alpha_{i,1}$, $\alpha_{i,2}$, and $\alpha_{i,3}$ listed in Table 3.14, each follower updates its position according to Eq. (2.17). Followers' desired configurations at four different sample times are shown in Fig. 3.40.

Table 3.14 Initial positions
of followers and
corresponding weights $\alpha_{i,k}$

i	X_i	Y_i	$\alpha_{i,1}$	$\alpha_{i,2}$	$\alpha_{i,3}$
4	11.5044	11.4005	0.6991	0.1608	0.1401
5	14.7762	10.5583	0.0448	0.8994	0.0558
6	14.3166	17.1550	0.1367	0.1478	0.7155
7	12.7596	13.4963	0.4481	0.2023	0.3496
8	13.1749	12.3376	0.3650	0.4012	0.2338
9	14.1295	11.6682	0.1741	0.6591	0.1668
10	14.3340	12.2003	0.1332	0.6468	0.2200
11	14.1000	14.6629	0.1800	0.3537	0.4663
12	13.4720	14.9644	0.3056	0.1980	0.4964
13	13.2542	14.1020	0.3492	0.2406	0.4102
14	13.5772	13.8027	0.2846	0.3352	0.3803

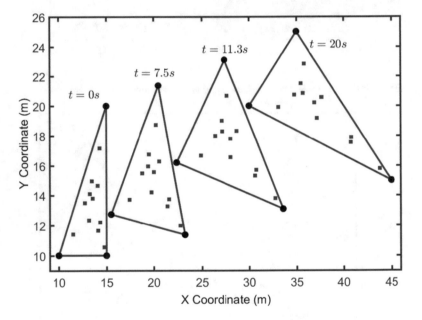

Fig. 3.40 Desired formations of the MAS at four different sample times

Followers' Evolution under Minimum Interagent Communication-Method 1:
Let the graph shown in Fig. 3.39 define fixed interagent communication among
the followers, where followers' communication weights are consistent with agents'
initial positions. Communication weights are specified by using Eq. (3.7) as listed
in Table 3.15. Let the follower $i \in V_F$ ($\forall i \in V_F$) update its current position according
to Eq. (3.21), where $g_i = g = 20$. In Fig. 3.41, components of the desired and actual
positions of follower 14 are shown by dotted and continuous curves, respectively.
As seen the follower 14 deviates from desired position $r_{i,HT}$ during evolution,
although the final desired position is reached. Followers' deviations from state of
homogeneous deformation is reduced by increasing the gain g.

Table 3.15 Initial positions and communication weights associated with the formation shown in Fig. 3.39

i	X_i	Y_i	i_1	i_2	i_3	w_{i,i_1}	w_{i,i_2}	w_{i,i_3}
1	10.0000	10.0000	–	–	–	–	–	–
2	15.0000	10.0000	–	–	–	–	–	–
3	15.0000	20.0000	–	–	–	–	–	–
4	11.5044	11.4005	1	7	8	0.50	0.20	0.30
5	14.7762	10.5583	2	9	10	0.71	0.15	0.14
6	14.3166	17.1550	3	11	12	0.45	0.25	0.30
7	12.7596	13.4963	4	12	13	0.32	0.30	0.38
8	13.1749	12.3376	4	9	14	0.29	0.36	0.35
9	14.1295	11.6682	5	8	10	0.35	0.31	0.34
10	14.3340	12.2003	5	9	11	0.33	0.37	0.30
11	14.1000	14.6629	6	10	14	0.40	0.30	0.30
12	13.4720	14.9644	6	7	13	0.34	0.29	0.37
13	13.2542	14.1020	7	12	14	0.35	0.35	0.30
14	13.5772	13.8027	8	11	13	0.30	0.41	0.29

Followers' Evolution under Minimum Interagent Communication-Method 2:
Now, let the follower i ($\forall i \in V_F$) uses the algorithm developed in Section 3.2.4, where the communication graph shown in Fig. 3.39 is applied. Thus, the follower $i \in V_F$ moves according to Eq. (3.124), where $g_i(t)$ is chosen as follows:

$$g_i(t) = 10 + max(|\lambda_k[(\dot{P}_i P_i^{-1})^T + \dot{P}_i P_i^{-1}]|) \tag{3.141}$$

Control parameter g_{14} is shown in Fig. 3.42 as a function of time. In addition, time varying weights $\varpi_{8,4}$, $\varpi_{8,9}$, and $\varpi_{8,14}$ are depicted versus time in Fig. 3.43. As seen in Fig. 3.43, final values of $\varpi_{8,4}$, $\varpi_{8,9}$, and $\varpi_{8,14}$ are the same as the communication weights $w_{8,4} = 0.29$, $w_{8,9} = 0.36$, and $w_{8,14} = 0.35$, respectively. This implies that the follower 14 ultimately reaches the desired position $r_{14,HT}$.

Followers' Evolution under Preservation of Area Ratios: Consider the initial distribution shown in Fig. 3.44 which is the same as agents' initial distribution in Fig. 3.39. As it is seen, followers apply a different communication graph to acquire the desired positions and each follower is allowed to interact with more than 3 local agents. In Table 3.16, agents' initial positions and the area weights, that are consistent with the graph shown in Fig. 3.44, are listed. Followers all choose control gain $g_i = g = 20$ ($\forall i \in V_F$), where they update their positions according to Eq. (3.140).

Transient area weights

$$aw_{ij} = \frac{a_{i,j}(t)}{a_i(t)}$$

are obtained as a function of time for the follower 14 as shown in Fig. 3.45. As seen in Fig. 3.45, the initial and final area weights are the same that conveys final formation is a homogeneous transformation of the initial distribution of the agents.

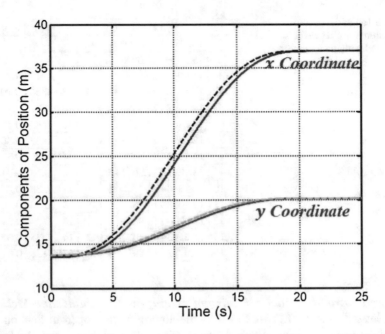

Fig. 3.41 X and Y components of the desired position $r_{14,HT}(t)$; X and Y components of the actual position $r_{14}(t)$

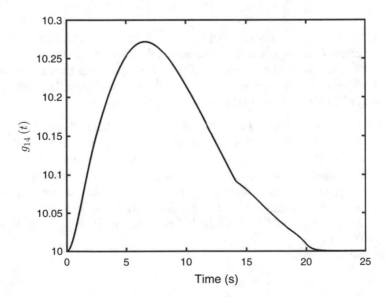

Fig. 3.42 Control parameter $g_{14}(t)$ in Example 3.9

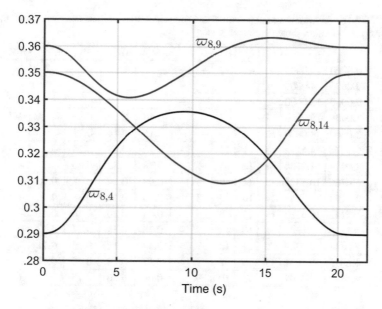

Fig. 3.43 Transient weight ratios $\varpi_{8,4}$, $\varpi_{8,9}$, and $\varpi_{8,14}$

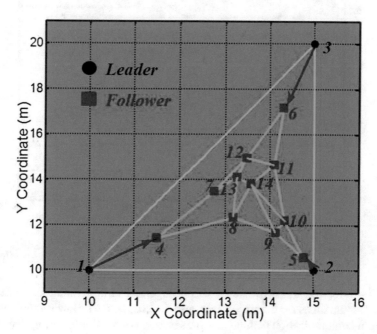

Fig. 3.44 The communication graph used for MAS evolution under the method of preservation of the volumetric ratios

Table 3.16 Followers' initial positions and the initial area weights associated with the MAS configuration shown in Figure 3.44

i	X_i	Y_i	i_1	i_2	i_3	i_4	i_5	AW_{i,i_1}	AW_{i,i_2}	AW_{i,i_3}	AW_{i,i_4}	AW_{i,i_5}
1	0.0000	0.0000	–	–	–	–	–	–	–	–	–	–
2	10.0000	4.0000	–	–	–	–	–	–	–	–	–	–
3	9.0000	12.0000	–	–	–	–	–	–	–	–	–	–
4	2.0000	15.0000	7	8	1	–	–	0.50	0.20	0.30	–	–
5	−3.0000	9.0000	10	2	9	–	–	0.15	0.14	0.71	–	–
6	1.8474	4.5152	3	11	12	–	–	0.30	0.45	0.25	–	–
7	7.6986	4.8704	4	12	13	–	–	0.38	0.32	0.30	–	–
8	6.6566	10.5475	13	14	9	4	–	0.08	0.24	0.28	0.40	–
9	2.8193	12.6743	14	10	5	8	–	0.20	0.16	0.18	0.46	–
10	1.3242	10.5533	14	11	5	9	–	0.43	0.20	0.16	0.21	–
11	−1.3275	8.8683	6	10	14	12	–	0.23	0.30	0.14	0.33	–
12	5.2544	7.0789	6	11	13	7	–	0.57	0.22	0.10	0.11	–
13	4.6015	9.3031	12	14	8	7	–	0.17	0.29	0.40	0.14	–
14	2.7536	9.1197	13	11	10	9	8	0.09	0.30	0.15	0.34	0.12

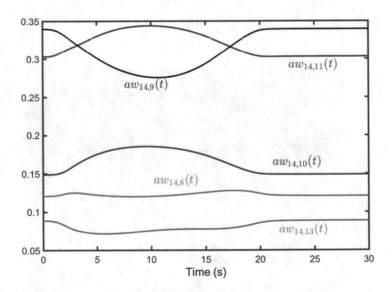

Fig. 3.45 Transient area weights of the follower 14 in Example 3.9

In Fig. 3.46, deviations from state of homogeneous transformation are depicted at three sample times $t = 6.5s$, $t = 11.3s$, and $t = 20s$, where followers apply the first minimum communication method and preservation of area ratios for updating their positions. As it is seen, deviation from the state of homogeneous transformation (no communication) is less when preservation of area ratios is applied.

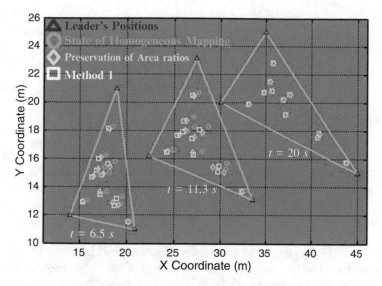

Fig. 3.46 Deviations from desired positions given by the homogeneous transformation in Example 3.9, where followers use the first method of minimum communication protocol and the method of preservation of area ratios

3.5   Discrete Time Dynamics for Homogeneous Deformation of an MAS [114]

Consider a collection of N agents moving in $\mathbb{R}^n$, where $n+1$ leaders are at the vertices of a leading convex polytope in $\mathbb{R}^n$, and followers are initially distributed inside the leading convex polytope. Let $r_i \in \mathbb{R}^n$ be updated by the following first order dynamics

$$r_{i+1}[K+1] = r_i[K] + u_i[K] \tag{3.142}$$

where

$$u_i[K] = \begin{cases} given & i \in V_L \\ r_i[K] + g \sum_{j \in N_i} w_{i,j} \left(r_j[K] - r_i[K] \right) & i \in V_F \end{cases}. \tag{3.143}$$

Therefore, dynamics of $x_{q,i}[K]$ (the qth component of position of the follower i) becomes

$$x_{q,i}[K+1] = x_{q,i}[K] + g \sum_{j \in N_i} w_{i,j} \left(x_{q,j}[K] - x_{q,i}[K] \right) \tag{3.144}$$

and the qth components of positions of the entire followers are updated by the following first order dynamics:

$$z_q[K+1] = z_q[K] + g\left(Az_q[K] + Bu_q[K]\right). \tag{3.145}$$

Notice that A and B are the partitions of the weight matrix W,

$$u_q = \left[x_{q,1}[K] \ \ldots \ x_{q,n+1}[K]\right]^T \in \mathbb{R}^{n+1}$$

$$z_q = \left[x_{q,n+2}[K] \ \ldots \ x_{q,N}[K]\right]^T \in \mathbb{R}^{N-n-1}$$

are the qth components of positions of the leaders and the followers, respectively. Because eigenvalues of F are all located inside the unit disk centered at the origin, the qth components of positions of followers asymptotically converge to

$$Z_{F,q} = -A^{-1}BU_{F,q}$$

where $U_{F,q} = U_q[K_F]$ denotes the qth components of final positions of the leaders at the time K_F. Note that the row $i - n - 1$ ($i \in V_F$) of Eq. (3.145) represents the qth component of ultimate position of the follower i, expressed by

$$K \geq K_F, \quad x_{q,i}[K] = \sum_{k=1}^{n+1} \alpha_{i,k} x_{q,k}[K_F] \tag{3.146}$$

as a convex combination of the qth components of the leaders' positions where the parameter $\alpha_{i,k}$ ($\forall i \in V_F, k \in V_L$) is uniquely determined by Eq. (2.15).

Following theorem provides an upper limit for deviations of followers' positions from the desired positions defined by a homogeneous transformation.

Theorem 3.8. *Let*

$$\forall K > 1 \quad and \quad k \in V_L, ||r_k[K+1] - r_k[K]|| \leq \gamma$$

then

$$\forall i \in V_F, ||r_i[K] - r_{i,HT}[K]|| \leq \delta \tag{3.147}$$

where

$$\delta = \frac{N-n-1}{1-\lambda_{max}}\gamma \tag{3.148}$$

specifies an upper bound for deviation of each follower $i \in V_F$ from the desired position $r_{i,HT}$ given by a homogeneous transformation, N is the total number of agents, and n is the dimension of the motion space. Eigenvalues of the matrix $I + gA$ are all located inside a disk with radius $\lambda_{max} < 1$ and center located at the origin [114].

Proof. Let Eq. (3.145) be rewritten as

$$
\begin{aligned}
z_q[K+1] &= z_q[K] + gA\left(z_q[K] - \left(-A^{-1}B\right)u_q[K]\right)\\
&= z_q[K] + gA\left(z_q[K] - z_{q,HT}[K]\right)
\end{aligned}
\tag{3.149}
$$

then

$$
E_q[K] = z_q[K] - z_{q,HT}[K]
$$

is updated by

$$
E_q[K+1] = (I+gA)E_q[K] - \left(z_{q,HT}[K+1] - z_{q,HT}[K]\right)
\tag{3.150}
$$

Therefore,

$$
\begin{aligned}
E_q[K] &= (I+gA)^K E_q[0]\\
&+ \sum_{l=0}^{K-1} (I+gA)^l \left(z_{q,HT}[K-l] - z_{q,HT}[K-1-l]\right)
\end{aligned}
\tag{3.151}
$$

Note that the first term in the right-hand side of Eq. (3.151) vanishes because $z_q[0] = z_{q,HT}[0] = W_L u_q[0]$.

Assume λ_{max} is the maximum eigenvalue of the matrix $I + gA$. Therefore,

$$
\begin{aligned}
\|E_q[K]\| &\le \sum_{l=0}^{K-1} \left(\lambda_{max}^l \mathbf{1}^T\right)\gamma\mathbf{1}\\
&= \gamma\frac{\mathbf{1}^T\mathbf{1}}{1-\lambda_{max}}
\end{aligned}
\tag{3.152}
$$

Substituting $\mathbf{1}^T\mathbf{1}$ by $N - n - 1$ results in

$$
\|E_q[K]\| \le \delta = \frac{N-n-1}{1-\lambda_{max}}\gamma.
$$

3.6 Final Remarks

3.6.1 *Homogeneous Deformation in a 3-D Space*

In this section homogeneous deformation of an MAS in a 3-D space is considered. The MAS consists of N agents (4 leaders and $N - 4$ followers), where leaders are at the vertices of a leading tetrahedron. Leaders' positions satisfy the following rank condition:

$$
\forall t \ge t_0,\ Rank\left[r_2 - r_1\ r_3 - r_1\ r_4 - r_1\right] = 3.
\tag{3.153}
$$

Homogeneous deformation in 3-D is expressed by

$$\begin{bmatrix} x_i(t) \\ y_i(t) \\ z_i(t) \end{bmatrix} = \begin{bmatrix} Q_{11}(t) & Q_{12}(t) & Q_{13}(t) \\ Q_{21}(t) & Q_{22}(t) & Q_{23}(t) \\ Q_{31}(t) & Q_{32}(t) & Q_{33}(t) \end{bmatrix} \begin{bmatrix} X_i \\ Y_i \\ Z_i \end{bmatrix} + \begin{bmatrix} D_1(t) \\ D_2(t) \\ D_3(t) \end{bmatrix}, \tag{3.154}$$

where $R_i = X_i\hat{\mathbf{e}}_x + Y_i\hat{\mathbf{e}}_y + Z_i\hat{\mathbf{e}}_z$ and $r_i(t) = x_i\hat{\mathbf{e}}_x + y_i\hat{\mathbf{e}}_y + z_i\hat{\mathbf{e}}_z$ denote the initial and current positions of the agent $i \in V$, respectively.

Because leaders' potions satisfy the rank condition (3.153), elements of the Jacobian matrix Q and rigid body displacement vector D can be uniquely related to the X, Y, and Z components of the leaders' positions at any time t. Let

$$J_t(t) = \begin{bmatrix} Q_{11} & Q_{12} & Q_{13} & Q_{21} & Q_{22} & Q_{23} & Q_{31} & Q_{32} & Q_{33} & D_1 & D_2 & D_3 \end{bmatrix}^T,$$

$$L_0 = \begin{bmatrix} X_1 & Y_1 & Z_1 \\ X_2 & Y_2 & Z_2 \\ X_3 & Y_3 & Z_3 \\ X_4 & Y_4 & Z_4 \end{bmatrix},$$

and

$$P_t = \begin{bmatrix} x_1 & x_2 & x_3 & x_4 & y_1 & y_2 & y_3 & y_4 & z_1 & z_2 & z_3 & z_4 \end{bmatrix}^T,$$

then

$$J_t(t) = \begin{bmatrix} I_3 \otimes L_0 & I_3 \otimes \mathbf{1} \end{bmatrix}^{-1} P_t,$$

where $I_3 \in \mathbb{R}^{3\times3}$ is the identity matrix and $\mathbf{1} \in \mathbb{R}^4$ is the one vector.

Followers are positioned inside the leading tetrahedron. Under a homogeneous deformation, positions of a follower $i \in V_F$ can be expressed by

$$r_i(t) = \sum_{k=1}^{s} \alpha_{i,k} r_k \tag{3.155}$$

where invariant parameters $\alpha_{i,1}$, $\alpha_{i,2}$, $\alpha_{i,3}$, and $\alpha_{i,4}$ are uniquely obtained from

$$\begin{bmatrix} X_1 & X_2 & X_3 & X_4 \\ Y_1 & Y_2 & Y_3 & Y_4 \\ Z_1 & Z_2 & Z_3 & Z_4 \\ 1 & 1 & 1 & 1 \end{bmatrix} \begin{bmatrix} \alpha_{i,1} \\ \alpha_{i,2} \\ \alpha_{i,3} \\ \alpha_{i,4} \end{bmatrix} = \begin{bmatrix} X_i \\ Y_i \\ Z_i \\ 1 \end{bmatrix}. \tag{3.156}$$

For example consider the initial distribution of the MAS shown in Fig. 3.47. Initial positions of the agents and parameters $\alpha_{i,1}$, $\alpha_{i,2}$, $\alpha_{i,3}$, and $\alpha_{i,4}$, that are consistent with agents' positions, are listed in Table 3.17.

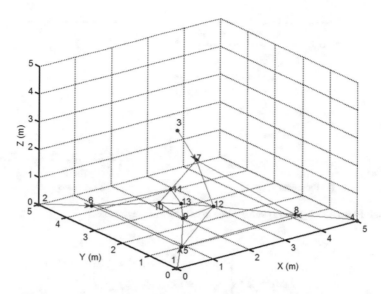

Fig. 3.47 Initial formation of an MAS moving as homogeneous deformation in a 3-*D* motion space. The MAS consists of four leaders and nine followers.

Table 3.17 Positions of agents in Fig. 3.47 and parameters $\alpha_{i,1}$, $\alpha_{i,2}$, $\alpha_{i,3}$, and $\alpha_{i,4}$

i	X_i	Y_i	Z_i	$\alpha_{i,1}$	$\alpha_{i,2}$	$\alpha_{i,3}$	$\alpha_{i,4}$
1	0	0	0	0	0	0	0
2	0	5.0000	0	0	0	0	0
3	0	0	5.0000	0	0	0	0
4	5.0000	0	0	0	0	0	0
5	0.5246	0.5275	0.3649	0.7166	0.1055	0.0730	0.1049
6	0.2576	3.4278	0.6069	0.1415	0.6856	0.1214	0.0515
7	0.6819	0.2246	3.5962	0.0995	0.0449	0.7192	0.1364
8	3.3938	0.1708	0.7095	0.1452	0.0342	0.1419	0.6788
9	0.7972	0.8382	1.1662	0.4397	0.1676	0.2332	0.1594
10	0.6654	1.5110	1.4665	0.2714	0.3022	0.2933	0.1331
11	0.6932	1.1197	2.1190	0.2136	0.2239	0.4238	0.1386
12	1.3659	0.4779	1.5668	0.3179	0.0956	0.3134	0.2732
13	0.8631	0.9886	1.5967	0.3103	0.1977	0.3193	0.1726

Leaders move along the trajectories defined by

$$Leader\ 1 : \begin{cases} x_1(t) = t & 0 \le t \le 20 \\ y_1(t) = 0 & 0 \le t \le 20 \\ z_1(t) = 0 & 0 \le t \le 20 \end{cases} \tag{3.157}$$

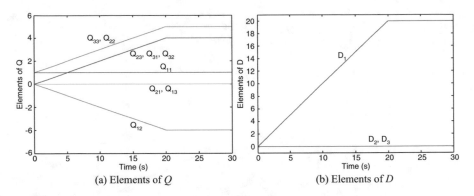

Fig. 3.48 Elements of Q and D versus time in example of Section 3.6.1

$$
\text{Leader 2}: \begin{cases} x_2(t) = 0 & 0 \le t \le 20 \\ y_2(t) = t+5 & 0 \le t \le 20 \\ z_2(t) = t & 0 \le t \le 20 \end{cases} \tag{3.158}
$$

$$
\text{Leader 3}: \begin{cases} x_3(t) = t & 0 \le t \le 20 \\ y_3(t) = t & 0 \le t \le 20 \\ z_3(t) = t+5 & 0 \le t \le 20 \end{cases} \tag{3.159}
$$

$$
\text{Leader 4}: \begin{cases} x_4(t) = t+5 & 0 \le t \le 20 \\ y_4(t) = 0 & 0 \le t \le 20 \\ z_4(t) = t & 0 \le t \le 20 \end{cases} \tag{3.160}
$$

and they stop at $(X_{1,F}, Y_{1,F}, Z_{1,F}) = (20,0,0)$, $(X_{2,F}, Y_{2,F}, Z_{2,F}) = (20,20,0)$, $(X_{3,F}, Y_{3,F}, Z_{3,F}) = (20,20,20)$, and $(X_{4,F}, Y_{4,F}, Z_{4,F}) = (20,0,20)$ at $t = 20s$. In Fig. 3.48, elements of the Jacobian matrix Q and the vector D are illustrated versus time. As it is seen, $Q(0) = I_3$ and $D(0) = 0 \in \mathbb{R}^3$.

Homogeneous Transformation under Local Communication: Followers can acquire desired homogeneous deformation through local communication. Let each follower agent i update its current position based on the positions of adjacent agents i_1, i_2, i_3, and i_4, where communication weights are uniquely determined based on the positions of the follower i and four in-neighbor agents as follows:

$$
\begin{bmatrix} X_{i_1} & X_{i_2} & X_{i_3} & X_{i_4} \\ Y_{i_1} & Y_{i_2} & Y_{i_3} & Y_{i_4} \\ Z_{i_1} & Z_{i_2} & Z_{i_3} & Z_{i_4} \\ 1 & 1 & 1 & 1 \end{bmatrix} \begin{bmatrix} w_{i,i_1} \\ w_{i,i_2} \\ w_{i,i_3} \\ w_{i,i_4} \end{bmatrix} = \begin{bmatrix} X_i \\ Y_i \\ Z_i \\ 1 \end{bmatrix}. \tag{3.161}
$$

Table 3.18 Positions of agents in Fig. 3.47 and parameters $\alpha_{i,1}$, $\alpha_{i,2}$, $\alpha_{i,3}$, and $\alpha_{i,4}$

i	i_1	i_2	i_3	i_4	w_{i,i_1}	w_{i,i_2}	w_{i,i_3}	w_{i,i_4}
5	1	6	8	9	0.6000	0.1000	0.1000	0.2000
6	2	5	10	11	0.6000	0.1000	0.1000	0.2000
7	3	8	11	12	0.6000	0.1000	0.1000	0.2000
8	4	5	7	12	0.6000	0.1000	0.1000	0.2000
9	5	10	12	13	0.3300	0.1500	0.1500	0.3700
10	6	9	11	13	0.2100	0.2200	0.3300	0.2400
11	6	10	7	13	0.1100	0.2200	0.3300	0.3400
12	7	8	5	13	0.2500	0.2500	0.2500	0.2500
13	9	10	11	12	0.2700	0.2200	0.2900	0.2200

It is noted that each follower is placed inside the ith ($i \in V_F$) communication tetrahedron whose vertices are occupied by the in-neighbor agents i_1, i_2, i_3, and i_4. Therefore, it is assured that communication weights are all positive.

Given communication weights, the weight matrix $W \in \mathbb{R}^{N-4 \times N}$ is set up by using Eq. (3.8) and partitions $A \in \mathbb{R}^{(N-4) \times (N-4)}$ and $B \in \mathbb{R}^{(N-4) \times 4}$ are determined according to Eq. (3.9). The matrix A is a Hurwitz matrix because communication weights are positive.

Each follower updates its position according to the first order dynamics (3.18) and (3.19). If time delay h_{1_i} and h_{2_i} are both zero, then the qth ($q = X, Y, Z$) components of the followers' positions are updated according to Eq. (3.22). Therefore, followers ultimately form a homogeneous deformation of the initial configuration.

Example 3.10. Consider an MAS with initial distribution shown in Fig. 3.47. Let leaders choose the trajectories given by Eqs. (3.158)–(3.160) and followers apply the communication graph shown in Fig. 3.47. Given initial positions of the agents (listed in Table 3.17), communication weights of the followers are specified by using Eq. (3.161). In Table 3.18, followers' in-neighbor agents and communication weights are given.

Shown in Fig. 3.49 are X, Y, and Z components of actual and desired positions of the follower 13. It is observed that follower 13 ultimately reaches the desired position prescribed by the homogeneous deformation.

Followers update their positions according to the first order dynamics (3.18) and (3.19), where $g_i = g = 20$ and communication delays are zero. In Fig. 3.50, formations of the MAS at $t = 0s$, $t = 5s$, $t = 10s$, and $t = 25s$ are shown by blue, green, red, and black, respectively.

3.6.2   *p-D Homogeneous Deformation in an n-D Space ($p \leq n$) [115]*

Let motion space be a linear motion space in $\mathbb{R}^n$ with orthonormal basis $(\hat{e}_1, \ldots, \hat{e}_n)$ and deformation space $M_p \subset \mathbb{R}^p$ ($p \leq n$) with orthonormal basis $(\tilde{e}_1, \ldots, \tilde{e}_p)$ be a

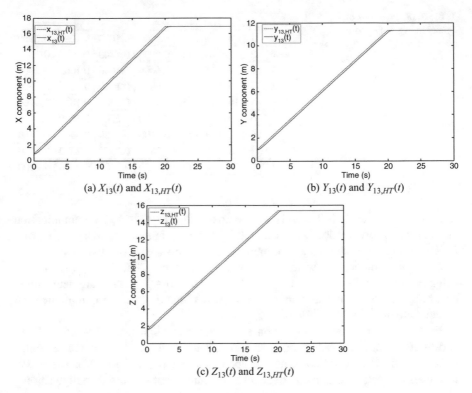

Fig. 3.49 X, Y, and Z components of actual and desired positions of the follower 13 in Example 3.10 **(a)** X components of actual and desired positions of follower 13 **(b)** Y components of actual and desired positions of follower 13 **(c)** Z components of actual and desired positions of follower 13

linear subspace of the motion space. Then $r_i \in \mathbb{R}^n$ (position of an agent $i \in V$ in the motion space) can be expressed by the following independent sum:

$$r_i = r' \oplus \tilde{r}_i \qquad (3.162)$$

where $r' \in \mathbb{R}^n$,

$$r_i = \sum_{q=1}^{n} x_{q,i} \hat{\mathbf{e}}_q,$$

and

$$\tilde{r}_i = \sum_{q=1}^{p} \tilde{x}_{q,i} \tilde{\mathbf{e}}_q.$$

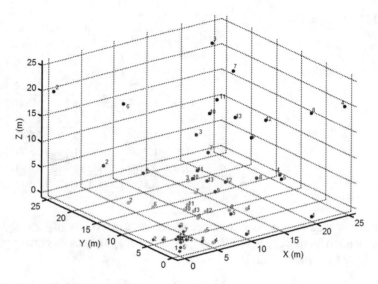

Fig. 3.50 Formations of the MAS at $t = 0s$, $t = 5s$, $t = 10s$, and $t = 25s$ in Example 3.11

Let positions of leaders 1, 2, ..., $p + 1$ ($V_L = \{1, 2, ..., p+1\}$) satisfy the rank condition

$$\forall t \geq t_0,\ Rank\ [r_2 - r_1\ ...\ r_{p+1} - r_1] = Rank\ [\tilde{r}_2 - \tilde{r}_1\ ...\ \tilde{r}_{p+1} - \tilde{r}_1] = p, \qquad (3.163)$$

then r_i ($i \in V = \{1, 2, ..., N\}$) can be expanded as

$$r_i(t) = \sum_{k=1}^{p+1} \tilde{p}_{i,k}(t)\,(r_k(t) - r_1(t)) \qquad (3.164)$$

where $\tilde{p}_{i,k}(t)$ is uniquely determined by solving the following set of $p + 1$ linear algebraic equations:

$$\begin{bmatrix} \tilde{x}_{1,1} & \tilde{x}_{1,2} & \cdots & \tilde{x}_{1,p+1} \\ \tilde{x}_{2,1} & \tilde{x}_{2,2} & \cdots & \tilde{x}_{2,p+1} \\ \vdots & \vdots & \ddots & \vdots \\ \tilde{x}_{p,1} & \tilde{x}_{p,2} & \cdots & \tilde{x}_{p,p+1} \\ 1 & 1 & \cdots & 1 \end{bmatrix} \begin{bmatrix} \tilde{p}_{i,1} \\ \tilde{p}_{i,2} \\ \vdots \\ \tilde{p}_{i,p} \\ \tilde{p}_{i,p+1} \end{bmatrix} = \begin{bmatrix} \tilde{x}_{1,i} \\ \tilde{x}_{2,i} \\ \vdots \\ \tilde{x}_{p,i} \\ 1 \end{bmatrix}. \qquad (3.165)$$

Under a homogeneous deformation, $\tilde{p}_{i,k}$ ($k = 1, 2, ..., p+1$) is time-invariant and is denoted by $\alpha_{i,k}$. Therefore, desired position given by a homogeneous deformation becomes

$$r_{i,HT}(t) = \sum_{k=1}^{p} \alpha_{i,k} r_k(t) \qquad (3.166)$$

where $\alpha_{i,k} = \tilde{p}_{i,k}(t_0)$ is obtained from Eq. (3.165).

Table 3.19 Initial positions of the agents in Example 3.11; in-neighbor agents and communication weights of the followers

i	X_i	Y_i	$\tilde{X}$	i_1	i_2	w_{i,i_1}	w_{i,i_2}
1	0.5000	−0.8660	1.0000	–	–	–	–
2	1.5000	−2.5981	3.0000	–	–	–	–
3	0.5735	−0.9933	1.1470	1	4	0.6100	0.3900
4	0.6885	−1.1925	1.3770	3	5	0.4100	0.5900
5	0.7684	−1.3309	1.5367	4	6	0.3900	0.6100
6	0.8195	−1.4193	1.6389	5	7	0.5700	0.4300
7	0.8872	−1.5366	1.7743	6	8	0.6900	0.3100
8	1.0379	−1.7977	2.0758	7	9	0.5500	0.4500
9	1.2221	−2.1167	2.4442	8	10	0.4300	0.5700
10	1.3610	−2.3574	2.7221	9	2	0.5000	0.5000

Followers acquire a desired homogeneous deformation in M_p through local communication. Follower $i \in V_F$ ($V_F = \{p+2, \ldots, N\}$) updates its current position according to the dynamics (3.21), where

$$r_{i,d} = \sum_{k=1}^{p+1} w_{i,i_k} r_{i,i_k}.$$

It is assumed that initial positions of the in-neighbor agents $i_1, i_2, \ldots, i_{p+1}$ satisfy the following rank condition:

$$Rank \left[R_{i_2} - R_{i_1} \, \cdots \, R_{i_{p+1}} - R_{i_1} \right] = Rank \left[\tilde{R}_{i_2} - \tilde{R}_{i_1} \, \cdots \, \tilde{R}_{i_{p+1}} - \tilde{R}_{i_1} \right] = p,$$
(3.167)

then communication weight w_{i,i_k} is the solution of

$$\begin{bmatrix} \tilde{X}_{1,i_1} & \tilde{X}_{1,i_2} & \cdots & \tilde{X}_{1,i_{p+1}} \\ \tilde{X}_{2,i_1} & \tilde{X}_{2,i_2} & \cdots & \tilde{X}_{2,i_{p+1}} \\ \vdots & \vdots & & \vdots \\ \tilde{X}_{p,i_1} & \tilde{X}_{p,i_2} & \cdots & \tilde{X}_{p,i_{p+1}} \\ 1 & 1 & & 1 \end{bmatrix} \begin{bmatrix} w_{i,i_1} \\ w_{i,i_2} \\ \vdots \\ w_{i,i_p} \\ w_{i,i_{p+1}} \end{bmatrix} = \begin{bmatrix} \tilde{X}_{1,i} \\ \tilde{X}_{2,i} \\ \vdots \\ \tilde{X}_{p,i} \\ 1 \end{bmatrix}.$$
(3.168)

It is noted that $\tilde{X}_{q,j} = \tilde{x}_{q,j}(t_0)$.

Example 3.11. Consider an MAS consisting of 10 agents with initial positions listed in Table 3.19. Agents 1 and 2, placed at opposite ends of the leading line segment, are the leaders and agents 3, ..., 10 are followers. It is noted that agents are initially distributed on the line segment makes the angle $\theta = \dfrac{-\pi}{3} rad$ with the X axis. Therefore, deformation subspace M_1 is one-dimensional.

Leaders move along trajectories defined as follows:

$$Leader\ 1: \begin{cases} x_1(t) = \cos(-\dfrac{\pi}{3} + \dfrac{5\pi}{60}t) & 0 \le t < 20 \\ y_1(t) = \sin(-\dfrac{\pi}{3} + \dfrac{5\pi}{60}t) & 0 \le t < 20 \end{cases}$$
(3.169)

$$
Leader\ 2: \begin{cases} x_2(t) = (3 - \dfrac{t}{20})\cos(-\dfrac{\pi}{3} + \dfrac{5\pi}{60}t) & 0 \le t < 20 \\[3mm] y_2(t) = (3 - \dfrac{t}{20})\sin(-\dfrac{\pi}{3} + \dfrac{5\pi}{60}t) & 0 \le t < 20 \end{cases}. \tag{3.170}
$$

Notice that leaders stop at $t = 20s$.

Given initial positions of the agents, communication weights of the followers are determined by

$$
\begin{cases} w_{i,i_1} = \dfrac{\tilde{X}_{i_2} - \tilde{X}_i}{\tilde{X}_{i_2} - \tilde{X}_{i_1}} \\[3mm] w_{i,i_2} = \dfrac{\tilde{X}_i - \tilde{X}_{i_1}}{\tilde{X}_{i_2} - \tilde{X}_{i_1}} \end{cases}. \tag{3.171}
$$

where i_1 and i_2 are index numbers of a follower i and

$$
\tilde{X}_i = (X_i\hat{\mathbf{e}}_x + Y_i\hat{\mathbf{e}}_y) \cdot \left(\cos\ (\dfrac{-\pi}{3})\hat{\mathbf{e}}_x + \sin\ (\dfrac{-\pi}{3})\hat{\mathbf{e}}_y \right). \tag{3.172}
$$

In Table 3.19 $\tilde{X}_i$, index numbers of the followers and followers' communication weights are also listed.

Followers acquire the desired 1-D homogeneous transformation in the $X - Y$ plane through local communication. Shown in Fig. 3.51 are formations of the MAS at different sample times.

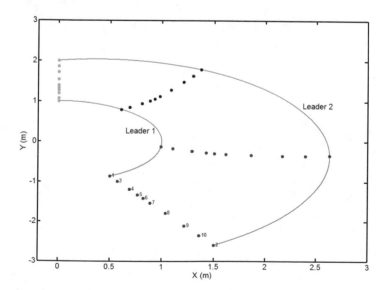

Fig. 3.51 Evolution of the MAS in Example 3.11

Chapter 4
Higher Order Dynamics for MAS Evolution as Continuum Deformation

In this chapter, it is demonstrated how followers can apply higher order dynamics to acquire a desired homogeneous deformation under local communication. First, evolution of followers using a second order dynamics with a nonlinear control gain g is considered. Then, asymptotic tracking of desired position issued by a homogeneous deformation is demonstrated, while followers applying higher order dynamics only communicate with their in-neighbor agents for updating their positions. Finally, evolution of followers in presence of heterogeneous communication delay is considered and upper bounds for followers' allowable communication delays are determined by using eigen-analysis.

4.1 Second Order Dynamics for Continuum Deformation of an MAS

Let the agent $i \in V$ update its position by

$$\begin{cases} \dot{r}_i = v_i \\ \dot{v}_i = a_i \end{cases} , \quad i \in V \tag{4.1}$$

with

$$a_i(t) = \begin{cases} \hat{u}_i(t) \in \mathbb{R}^n & i \in V_L \\ -c_i v_i(t) + g_i \sum_{j \in N_i} w_{i,j}(r_j(t) - r_i(t)) & i \in V_F \end{cases} \tag{4.2}$$

© Springer International Publishing AG 2016
H. Rastgoftar, *Continuum Deformation of Multi-Agent Systems*,
DOI 10.1007/978-3-319-41594-9_4

where $\hat{u}_i(t)$ is the acceleration of the leader $i \in V_L$ that is known at the time t. Let

$$g_i = g_{0i} + g_{2i}\|r_{i,d} - r_i\|^2 \tag{4.3}$$

with $g_{0i} \in \mathbb{R}_+$ and $g_{2i} \in \mathbb{R}_+$, then g_i has a local minimum at $r_{i,d}$. The qth component of Eq. (4.1) is the row $i - n - 1$ of the following second order matrix dynamics representing evolution of the qth components of the positions of the followers:

$$\ddot{z}_q + C\dot{z}_q - GAz_q = GBu_q. \tag{4.4}$$

It is noted that $z_q(t) = [x_{q,n+2} \cdots x_{q,N}]^T \in \mathbb{R}^{N-n-1}$ and $u_q(t) = [x_{q,1} \cdots x_{q,n+1}]^T \in \mathbb{R}^{n+1}$ are the qth components of the positions of the followers and leaders, respectively. Also, $A \in \mathbb{R}^{(N-n-1)\times(N-n-1)}$ and $B \in \mathbb{R}^{(N-n-1)\times(n+1)}$ are partitions of the weight matrix W (See Section 3.2.), $C = diag(c_{n+2}, \ldots, c_N) \in \mathbb{R}^{(N-n-1)\times(N-n-1)}$ and $G = diag(g_{n+2}, \ldots, g_N) \in \mathbb{R}^{(N-n-1)\times(N-n-1)}$ are positive diagonal matrices. Because matrix A is Hurwitz, the dynamics of evolution of the followers, given by Eq. (4.4), is stable. Notice that the equilibrium state of Eq. (4.4) satisfies Eq. (3.15), thus the final formation of the agents is a homogeneous transformation of the initial configuration.

Example 4.1. Consider an MAS consisting of 10 agents (3 leaders and 7 followers), moves in a plane. Every follower is a unicycle-like robot, where the centroid position of each robot is updated through local communication according to the dynamics (4.1) and (4.2). Schematic of a unicycle-like follower is illustrated in Fig. 4.1. As it is seen, each unicycle-like robot is considered as a 0.30 *m* by 0.30 *m* square, where the center of mass of every follower robot is located at the center

Fig. 4.1 Schematic of a follower unicycle-like robot

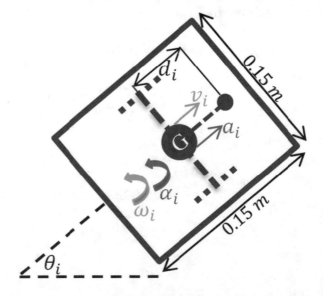

of the square, and $d_i = 0.1m$. Let $r_i = x_i\hat{e}_x + y_i\hat{e}_y$ denote centroid position of the follower i, then linear velocity v_i, angular velocity $\omega_i = \dot{\theta}_i$, acceleration a_i, and angular acceleration $\alpha_i = \dot{\omega}_i$ can be uniquely related to the X and Y components of the centroid velocity and acceleration of the follower i as follows:

$$\begin{bmatrix} \dot{x}_i \\ \dot{y}_i \end{bmatrix} = \begin{bmatrix} \cos\theta_i & -d_i\sin\theta_i \\ \sin\theta_i & d_i\cos\theta_i \end{bmatrix} \begin{bmatrix} v_i \\ \omega_i \end{bmatrix} \tag{4.5}$$

$$\begin{bmatrix} \ddot{x}_i \\ \ddot{y}_i \end{bmatrix} = \begin{bmatrix} \cos\theta_i & -d_i\sin\theta_i \\ \sin\theta_i & d_i\cos\theta_i \end{bmatrix} \begin{bmatrix} a_i \\ \alpha_i \end{bmatrix} - d_i\omega_i^2 \cdot \begin{bmatrix} \cos\theta_i \\ \sin\theta_i \end{bmatrix}. \tag{4.6}$$

This implies that v_i, ω_i, a_i, α_i can be formulated based on the centroid velocity and acceleration as follows:

$$\begin{bmatrix} v_i \\ \omega_i \end{bmatrix} = \begin{bmatrix} \cos\theta_i & -d_i\sin\theta_i \\ \sin\theta_i & d_i\cos\theta_i \end{bmatrix}^{-1} \begin{bmatrix} \dot{x}_i \\ \dot{y}_i \end{bmatrix} \tag{4.7}$$

$$\begin{bmatrix} a_i \\ \alpha_i \end{bmatrix} = \begin{bmatrix} \cos\theta_i & -d_i\sin\theta_i \\ \sin\theta_i & d_i\cos\theta_i \end{bmatrix}^{-1} \begin{bmatrix} \ddot{x}_i + d_i\omega_i^2\cos\theta_i \\ \ddot{y}_i + d_i\omega_i^2\sin\theta_i \end{bmatrix}. \tag{4.8}$$

Note that a_i and α_i are the control inputs executed by each follower robot. In this regard, first the follower agent $i \in V_F$ determines the X and Y components of its centroid acceleration through communication with three local agents according to Eqs. (4.1) and (4.2), where $g = g_i = g_{0i} = 65$ ($\forall i \in V_F, g_{2i} = 0$). Then the centroid velocity and acceleration of the follower i are related to v_i, ω_i, a_i, and α_i by using Eqs. (4.7) and (4.8).

Initial positions of the agents and the graph defining interagent communication among the agents are shown in Fig. 4.2. Initial centroid positions of the followers and leaders as well as followers' initial heading angles are listed in Table 4.1. Additionally communication weights that are consistent with agents' initial positions and the graph shown in Fig. 4.2 are listed in the last three columns of Table 4.1.

Scenario I (Rigid Body Translation). Suppose that leaders move with the same velocity on three parallel straight paths shown in Fig. 4.3(a). Leaders start their motion from the rest at $t = 0$, and they finally stop at $(X_{1,F}, Y_{2,F}) = (5, 14)$, $(X_{2,F}, Y_{2,F}) = (15, 14)$, and $(X_{3,F}, Y_{3,F}) = (9, 24)$ at $T = 60s$. Furthermore, X component of the leaders' acceleration (q_1, q_2, and q_3) and Y component of the leaders' acceleration (q_4, q_5, and q_6) are depicted versus time in Fig. 4.3(b).

Formations of the MAS at five different sample times $t = 5s$, $t = 10s$, $t = 15s$, $t = 30s$, and $t = 75s$ are shown in Fig. 4.4. In Fig. 4.5 heading angles of follower robots are shown versus time. As it is seen, followers' final heading angles converge to a same value because leaders move with same velocities on parallel straight paths.

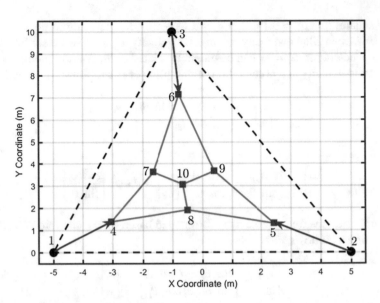

Fig. 4.2 Communication graph used in Example 4.1

Table 4.1 Followers' initial heading angles, agents' initial positions, and communication weights in Example 4.1

i	θ_i (deg)	X_i	Y_i	i_1	i_2	i_3	w_{i,i_1}	w_{i,i_2}	w_{i,i_3}
1	-	−5	0	-	-	-	-	-	-
2	-	5	0	-	-	-	-	-	-
3	-	1	10	-	-	-	-	-	-
4	30	−3.0400	1.4058	1	7	8	0.50	0.25	0.25
5	60	2.4228	1.3506	2	8	9	0.50	0.30	0.20
6	210	−0.7862	7.1847	3	7	9	0.55	0.20	0.25
7	300	−1.6516	3.6612	4	6	10	0.40	0.30	0.30
8	120	−0.5085	1.9621	4	5	10	0.35	0.32	0.33
9	0	0.3767	3.8096	5	6	10	0.35	0.32	0.33
10	0	−0.6659	3.1450	7	8	9	0.37	0.33	0.30

In Fig. 4.6 X and Y components of the desired centroid position of the follower 10 given by the homogeneous transformation are shown by dotted curves. Moreover, X and Y components of actual centroid position of the follower 10 are depicted by continuous curves.

Scenario II. For this example, agents' initial positions and the corresponding communication weights are the same as Table 3.15, but initial heading angles of the followers are all zero ($\theta_i(0) = 0, \forall i \in V_F$). Let leaders choose the optimal trajectories (minimizing the acceleration norm of MAS evolution) specified in Example 2.1, and

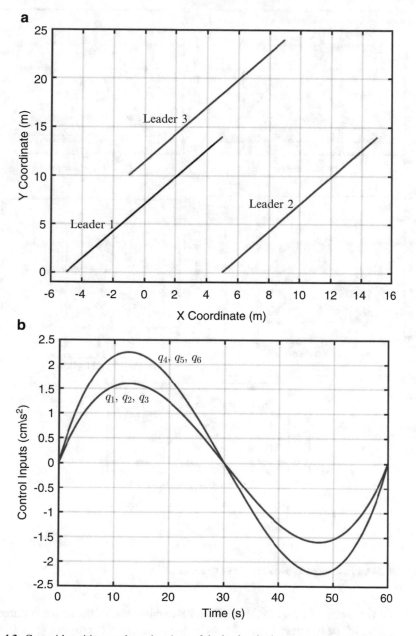

Fig. 4.3 Centroid positions and accelerations of the leaders in the scenario I of Example 4.1

centroid positions of the followers be updated according to the dynamics (4.1) and (4.2). Then, the control inputs a_i, and α_i are obtained by using Eqs. (4.7) and (4.8).

Shown in Fig. 4.7 are formations of the MAS at five different sample times $t = 5s$, $t = 10s$, $t = 15s$, $t = 30s$, and $t = 70s$. Furthermore, the followers' heading angles are

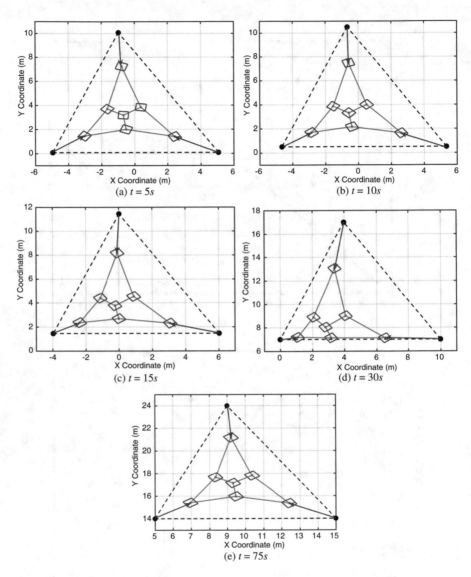

Fig. 4.4 MAS formations at five different sample times

depicted versus time in Fig. 4.8. Also, X and Y components of the actual position of the centroid of the follower 10 are shown by the continuous curves in Fig. 4.9. Note that the X and Y components of the desired position of the centroid of the follower 10 are shown by the dotted curves in Fig. 4.9.

Example 4.2 (Nonlinear control gain g_i). Consider the MAS consisting of 20 agents with the agents' initial formation shown in Fig. 2.5, where leaders are initially $(X_1, Y_1) = (-5, 0)$, $(X_2, Y_2) = (5, 0)$, and $(X_3, Y_3) = (-1, 10)$. Leaders 1, 2, and

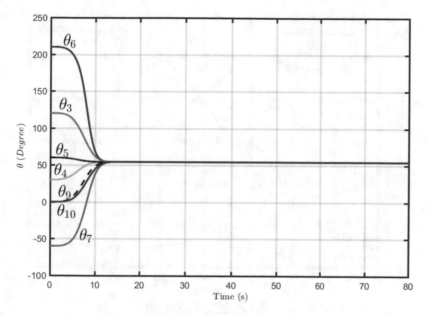

Fig. 4.5 Orientations of follower robots versus time in the scenario I of Example 4.1

3 choose the paths shown in Fig. 2.6 to reach the desired final destinations at $(X_{1,F}, Y_{1,F}) = (6, -4)$, $(X_{2,F}, Y_{2,F}) = (14, 16)$, and $(X_{3,F}, Y_{3,F}) = (6, 8.5)$. Elements of the Jacobian Q and the vector D corresponding to the leaders' positions are shown versus time in Fig. 4.10.

Notice that leaders' accelerations are chosen such that the cost function (2.65) is minimized. Also, leaders' positions satisfy constraint Eq. (2.63). Therefore, the acceleration norm of homogeneous transformation is minimized, where the area of the leading triangle is preserved during MAS evolution. The communication weights applied by the followers are obtained by using Eq. (3.7) based on the initial positions of the agents shown in Fig. 2.5, as listed in Table 4.2. Followers update their positions according to the dynamics (4.1) and (4.2). Configurations of the agents at four different times $t = 15s$, $t = 30s$, $t = 50s$, and $t = 75s$ are illustrated in Fig. 4.11.

Force Analysis: When followers acquire the desired positions through local communication, the required force may exceed the bound, $\sqrt{a_1^2 + a_2^2}$ (See the relation (2.41)). This is because of the time lag of followers in acquiring desired positions prescribed by the leaders. In Fig. 4.12, the required forces for the followers are shown. As observed, the acting forces required for the motion of the followers exceed the upper bound $\sqrt{a_1^2 + a_2^2}$ that was previously determined in Example 2.2.

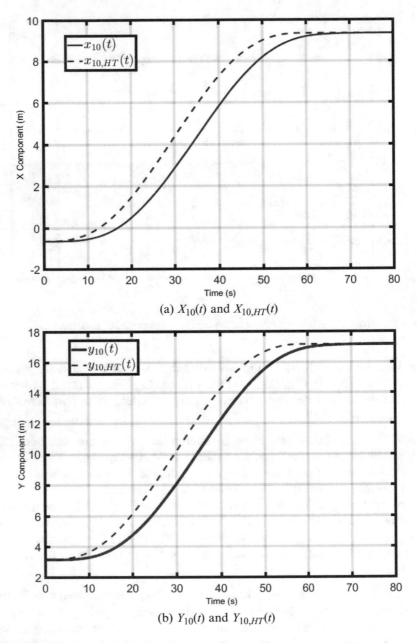

(a) $X_{10}(t)$ and $X_{10,HT}(t)$

(b) $Y_{10}(t)$ and $Y_{10,HT}(t)$

Fig. 4.6 X and Y components of position of the follower 10 in the scenario I of Example 4.1

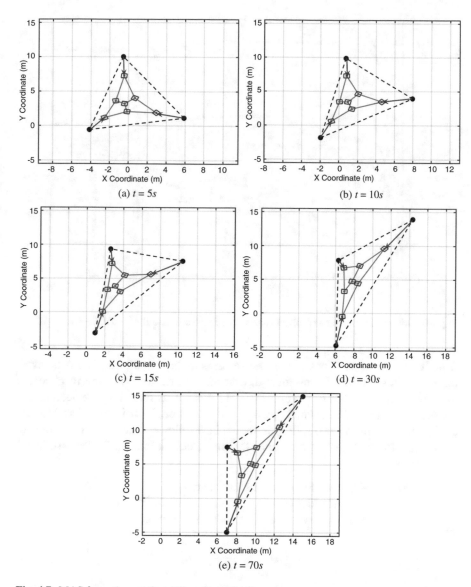

Fig. 4.7 MAS formations at five different sample times

4.2 Asymptotic Tracking of Desired Positions [115]

The minimum communication protocol, presented in Chapter 3, can assure convergence of followers' positions in the current configuration to the desired positions inside the final containment region, when leaders stopped. However, followers deviate from the desired positions, defined by homogeneous transformation, during transition. If followers' deviations are considerable, interagent collision may not be necessarily avoided during transition. Therefore, vanishing followers' deviations

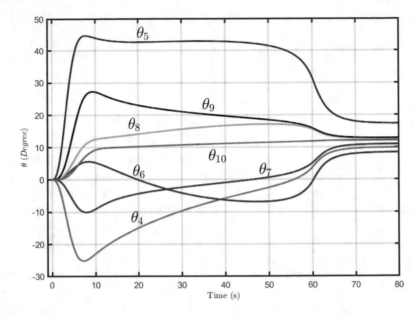

Fig. 4.8 Orientations of follower robots versus time in the scenario II of Example 4.1

from their desired positions during transition is highly significant, and it is investigated in this section. In this regard, first each leader is permitted to choose a finite polynomial vector of the order $(p-1) \in \mathbb{N}$ for its trajectory connecting two consecutive way points. Then, a dynamics of the order $p \in \mathbb{N}$ is considered for the evolution of the follower $i \in V_F$. It is shown how followers' deviations converge to zero during transition, while each follower only access the state information of $n+1$ local agents.

Dynamics of the Agents: Suppose that the leader agent k $(k \in V_L)$ chooses a finite polynomial vector of the order $(p-1)$ $(\in \mathbb{N})$, as a function of time, for its trajectory connecting two desired consecutive way points. Therefore,

$$\forall k \in V_L, \quad \frac{d^p r_k}{dt^p} = 0. \tag{4.9}$$

Follower i $(i \in V_F)$ updates its current position by

$$\frac{d^p r_i}{dt^p} = \beta_1 \left(\frac{d^{p-1} r_{i,d}}{dt^{p-1}} - \frac{d^{p-1} r_i}{dt^{p-1}} \right) + \cdots + \beta_p \left(r_{i,d} - r_i \right) \tag{4.10}$$

where $r_i \in \mathbb{R}^n$ is the actual position of follower i,

$$r_{i,d}(t) = \sum_{q=1}^{n} x_{q,i,d}(t) \hat{\mathbf{e}}_q = \sum_{j \in N_i} w_{i,j} x_j(t), \tag{4.11}$$

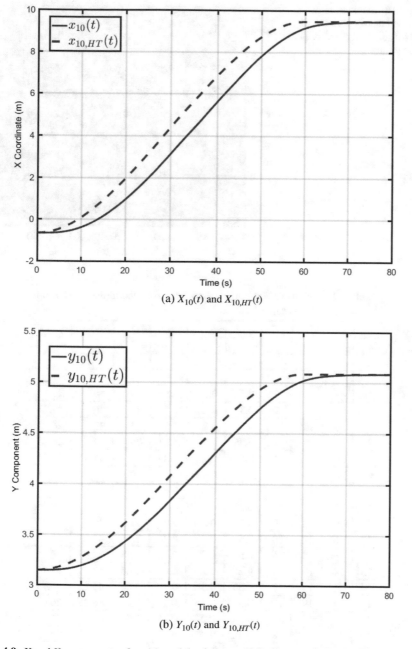

(a) $X_{10}(t)$ and $X_{10,HT}(t)$

(b) $Y_{10}(t)$ and $Y_{10,HT}(t)$

Fig. 4.9 X and Y components of position of the follower 10 in the scenario II of Example 4.1

and the constant parameters β_1, β_2, $\ldots$, β_p are chosen such that the roots of the characteristic equation

$$|s^p I - (\beta_1 s^{p-1} + \cdots + \beta_{p-1} s + \beta_p)A| = 0 \qquad (4.12)$$

Fig. 4.10 Elements of the Jacobian matrix Q and rigid body displacement vector D versus time

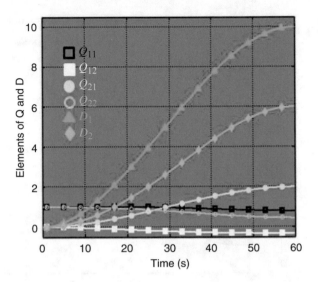

Table 4.2 Communication weights associated with the formation shown in Fig. 2.5; parameters g_{0i}, g_{2i}, and c_i

i	i_1	i_2	i_3	w_{i,i_1}	w_{i,i_2}	w_{i,i_3}	g_{0i}	g_{2i}	c_i
1	-	-	-	-	-	-	-	-	-
2	-	-	-	-	-	-	-	-	-
3	-	-	-	-	-	-	-	-	-
4	1	7	8	0.3	0.2	0.5	70	45	25.5
5	2	8	9	0.3	0.2	0.5	72	46	27.25
6	3	13	14	0.3	0.2	0.5	74	47	26.75
7	4	8	11	0.55	0.2	0.25	76	48	26.25
8	5	7	11	0.6	0.2	0.2	69	44.5	29.25
9	5	10	12	0.55	0.25	0.2	78	49	27.75
10	9	11	19	0.35	0.23	0.42	73	46.5	27.5
11	7	8	10	0.33	0.2	0.47	80	50	28
12	9	13	20	0.5	0.3	0.2	71	45.5	28.25
13	6	12	20	0.5	0.3	0.2	72.5	46.25	25.5
14	6	15	16	0.55	0.27	0.18	69.5	44.75	30
15	14	16	18	0.2	0.2	0.6	75.5	47.75	25.75
16	14	15	17	0.35	0.36	0.29	78.5	49.25	29.75
17	16	19	20	0.3	0.35	0.35	73.5	46.75	26
18	4	15	19	0.55	0.2	0.25	71.5	45.75	27.75
19	10	17	18	0.32	0.42	0.26	77.5	48.75	29.75
20	12	13	17	$\frac{1}{3}$	$\frac{1}{3}$	$\frac{1}{3}$	79	49.5	26.5

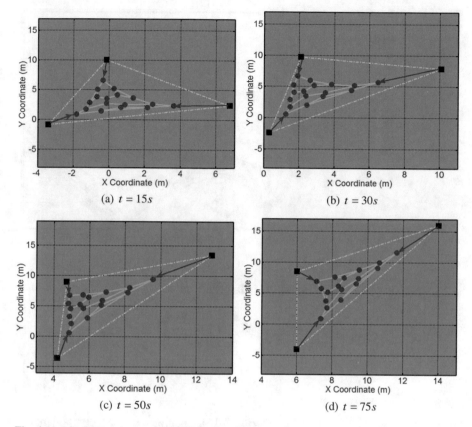

Fig. 4.11 MAS formations at five different sample times

are all located in the open left half s-plane. Therefore, $x_{q,i}$ (the qth component of the position of the follower i) is updated as follows:

$$\frac{d^p x_{q,i}}{dt^p} = \beta_1 \left(\frac{d^{p-1} x_{q,i,d}}{dt^{p-1}} - \frac{d^{p-1} x_{q,i}}{dt^{p-1}} \right) + \cdots + \beta_p (x_{q,i,d} - x_{q,i}) \tag{4.13}$$

where $x_{q,i,d}(t)$ is the qth component of $r_{i,d}(t)$. Equation (4.13) is the row $i - n - 1$ of the following matrix dynamics of the order p:

$$\frac{d^p z_q}{dt^p} - A \left(\beta_1 \frac{d^{p-1} z_q}{dt^{p-1}} + \beta_2 \frac{d^{p-2} z_q}{dt^{p-2}} + \cdots + \beta_p z_q \right) = B \left(\beta_1 \frac{d^{p-1} u_q}{dt^{p-1}} + \beta_2 \frac{d^{p-2} u_q}{dt^{p-2}} + \cdots + \beta_p u_q \right).$$

$$\tag{4.14}$$

Fig. 4.12 Magnitude of control forces executed by followers when they update their positions through local communication

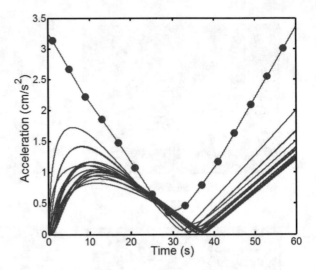

Now the right-hand side of Eq. (4.14) is rewritten as

$$B\left(\beta_1\frac{d^{p-1}u_q}{dt^{p-1}}+\beta_2\frac{d^{p-2}u_q}{dt^{p-2}}+\cdots+\beta_p u_q\right)=AA^{-1}B\left(\beta_1\frac{d^{p-1}u_q}{dt^{p-1}}+\beta_2\frac{d^{p-2}u_q}{dt^{p-2}}+\cdots+\beta_p u_q\right)=$$
$$-A\left(\beta_1\frac{d^{p-1}z_{q,HT}}{dt^{p-1}}+\beta_2\frac{d^{p-2}z_{q,HT}}{dt^{p-2}}+\cdots+\beta_p z_{q,HT}\right).$$
$$(4.15)$$

where $z_{q,HT}\in\mathbb{R}^{N-n-1}$, defined by Eq. (3.15), specifies the qth components of followers' desired positions at time t. Then, Eq. (4.14) is simplified to

$$\frac{d^p E_q}{dt^p}-A\left(\beta_1\frac{d^{p-1}E_q}{dt^{p-1}}+\beta_2\frac{d^{p-2}E_q}{dt^{p-2}}+\cdots+\beta_p E_q\right)=\frac{d^p z_{q,HT}}{dt^p} \qquad (4.16)$$

where $E_q=z_{q,HT}-z_q$ denotes the qth component of deviation of followers from the desired state defined by a homogeneous transformation.

Because leaders' positions satisfy Eq. (4.9), thus $\dfrac{d^p z_{q,HT}}{dt^p}=0$ and E_q asymptotically converges to zero. In other words, deviations of followers from desired positions vanish during MAS evolution, while followers can only access states of $n+1$ local agents.

Example 4.3. Consider an MAS consisting of 3 leaders and 11 followers evolves in the $X-Y$ plane. Leaders move with constant velocities on the straight paths illustrated in Fig. 4.13. Therefore,

$$\frac{d^2 r_k}{dt^2}=0,\ k=1,2,3.$$

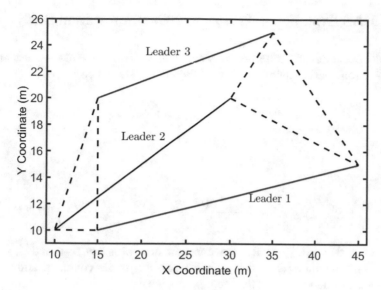

Fig. 4.13 Paths of leaders in Example 4.3

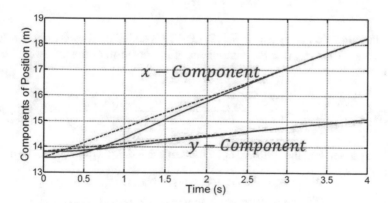

Fig. 4.14 Asymptotic tracking of desired position by the follower 14

Each follower updates the qth component of its current position according to the dynamics (4.10) based on the positions of three adjacent agents, where $p = 2$, $\beta_1 = 20$, and $\beta_2 = 20$. Initial positions of the agents are shown in Fig. 3.44 as listed in Table 3.16. Shown in Fig. 4.14 are the X and Y components of the desired and actual positions of the follower 14 as a function of time. As it is seen, the desired position shown by dotted curve is asymptotically tracked by the follower 14 during MAS evolution and deviation of follower 14 (from its desired position) vanishes in around 3s.

4.3   MAS Evolution in Presence of Communication Delay

Consider evolution of a collection of N agents in $\mathbb{R}^n$ consisting of $n+1$ leaders and $N - n - 1$ followers. Suppose that each follower updates its position by

$$\frac{d^2 r_i}{dt^2} = u_i, \ i \in V_F \tag{4.17}$$

where

$$u_i = \begin{cases} 0 & i \in V_L \\ \sum_{k=1}^{2} \sum_{j \in N_i} g_k^{\ i} w_{i,j} \left(\frac{d^{2-k} r_j(t - h_{i,k})}{dt^{2-k}} - \frac{d^{2-k} r_i(t - h_{i,k})}{dt^{2-k}} \right) & i \in V_F \end{cases}.$$

$$\tag{4.18}$$

It is noted that $g_1^{\ i}, g_2^{\ i} \in \mathbb{R}_+$ are constant control gains, $h_{i,1}$ and $h_{i,2} \in \mathbb{R}_+$ are constant communication delays, and $w_{i,j} \in \mathbb{R}_+$ is the communication weight between the agent $i \in V_F$ and the agent $j \in N_i$, where

$$\sum_{j \in N_i} w_{i,j} = 1.$$

The communication delay $h_{i,k}$ ($k = 1, 2, i \in V_F$) is called the *k-delay of the follower i*. The dynamics of $x_{q,i}$ (the qth component of the position of the follower i) is then becomes

$$\frac{d^2 x_{q,i}}{dt^2} - \sum_{k=1}^{2} \sum_{j \in N_i} g_k^{\ i} w_{i,j} \left(\frac{d^{2-k} x_{q,j}(t - h_{i,k})}{dt^{2-k}} - \frac{d^{2-k} x_{q,i}(t - h_{i,k})}{dt^{2-k}} \right) = 0. \tag{4.19}$$

It is supposed that the follower i has the same $k - delay$ $h_{i,k}$ ($k = 1, 2$) to measure $\frac{d^{2-k} x_{q,j}(t - h_{i,k})}{dt^{2-k}}$ of the agent $j \in N_i$. Let $\hat{z}_q(s) = [\hat{x}_{q,n+2}(s) \ \ldots \hat{x}_{q,N}(s)]^T$ and $\hat{u}_q(s) = [\hat{x}_{q,1}(s) \ \ldots \hat{x}_{q,n+1}(s)]^T$ be the Laplace transform of $z_q(t) = [x_{q,n+2}(t)(t) \ \ldots x_{q,N}(t)]^T$ and $u_q(t) = [x_{q,1}(t) \ \ldots x_{q,n+1}(t)]^T$, respectively, then the characteristic equation of MAS evolution dynamics becomes

$$|s^2 I - G_1 T_1 s + G_2 T_2) A| = 0, \tag{4.20}$$

where

$$T_k(s) = diag(e^{-h_{n+2,k} s}, \ldots, e^{-h_{N,k} s}) \in \mathbb{R}^{(N-n-1) \times (N-n-1)}, \ k = 1, 2 \tag{4.21}$$

$$G_k = diag(g_k^{\ n+2}, \ldots, g_k^{\ N}) \in \mathbb{R}^{(N-n-1) \times (N-n-1)}, \ k = 1, 2 \tag{4.22}$$

are diagonal matrices. To assure stability of the MAS evolution, the roots of Eq. (4.20) are required to be located in the open left half s-plane. Note that the roots of Eq. (4.20) are the same as the roots of the following characteristic equation:

$$\Phi(s,A) = |\pi(s) - A| = 0, \tag{4.23}$$

where

$$\pi(s) = diag(\pi_4, \ldots, \pi_N) \in \mathbb{R}^{(N-n-1) \times (N-n-1)} \tag{4.24}$$

and

$$\pi_i(s, h_{i,1}, h_{i,2}) = \frac{s^2}{g_1{}^i s e^{-h_{i,1}s} + g_2{}^i e^{-h_{i,2}s}}, \; i \in V_F. \tag{4.25}$$

At the margin of instability, s crosses $j\omega$ axis, therefore, upper bounds for communication delays $h_{i,1}$ and $h_{i,2}$ are obtained by analysis of

$$\Phi(j\omega, A) = 0. \tag{4.26}$$

Therefore,

$$\Phi_{i,l}(j\omega_{i,l}, h_{i,1}{}^l, h_{i,2}{}^l, \lambda_l) = \pi(j\omega, h_{i,1}^l, h_{i,2}^l) - \lambda_l = 0. \tag{4.27}$$

Equation (4.27) can be rewritten in the following component wise form:

$$\begin{bmatrix} \sin(\omega_{i,l}h_{i,1}^l - \theta_l) & \cos(\omega_{i,l}h_{i,2}^l - \theta_l) \\ \cos(\omega_{i,l}h_{i,1}^l - \theta_l) & -\sin(\omega_{i,l}h_{i,2}^l - \theta_l) \end{bmatrix} \begin{bmatrix} g_1^i a_l \omega_{i,l} \\ g_2^i a_l \end{bmatrix} = \begin{bmatrix} -\omega_{i,l}^2 \\ 0 \end{bmatrix}. \tag{4.28}$$

Let

$$H_i = \{(h_{i,1}^l, h_{i,2}^l) | \Phi_{i,l}(j\omega_{i,l}, h_{i,1}^l, h_{i,2}^l, \lambda_l) = 0, h_{i,1}^l > 0, h_{i,2}^l > 0\} = 0, \; l = 1, 2, \ldots, N-n-1, \tag{4.29}$$

then,

$$(h_{i,1}, h_{i,2}) = argmin_{(h_{i,1}^l, h_{i,2}^l) \in H_i} \sqrt{h_{i,1}^l{}^2 + h_{i,2}^l{}^2} \tag{4.30}$$

specifies an upper bound for the communication delays $h_{i,1}^l$ and $h_{i,2}^l$.

Existence of $\omega_{i,l}$, $h_{i,1}^l$ and $h_{i,2}^l$ for $\lambda_l \in \mathbb{R}$ when eigenvalues of A are real: If eigenvalues of A are all real, then $\theta_l = \pi$ and the communication delays $h_{i,1}^l$ and $h_{i,2}^l$ in Eq. (4.28) are obtained as follows:

$$h_{i,1}^l = \frac{1}{\omega_{i,l}}\left(\frac{\pi}{2} - \cos^{-1}\left(\frac{\left(g_2^i a_l\right)^2 - \left(g_1^i a_l \omega_{i,l}\right)^2 - \left(\omega_{i,l}\right)^4}{2 g_1^i a_l \omega_{i,l}^3}\right)\right) \tag{4.31}$$

$$h_{i,2}^l = \frac{1}{\omega_{i,l}}\cos^{-1}\left(\frac{\left(g_1^i a_l \omega_{i,l}\right)^2 - \left(g_2^i a_l\right)^2 - \left(\omega_{i,l}\right)^4}{2 g_2^i a_l \omega_{i,l}^2}\right) \tag{4.32}$$

Note that $h_{i,1}$ and $h_{i,2}$ positive, if $\omega_{i,l}$ satisfies the following equations:

$$0 < \frac{\left(g_2^i a_l\right)^2 - \left(g_1^i a_l \omega_{i,l}\right)^2 - \left(\omega_{i,l}\right)^4}{2 g_1^i a_l \omega_{i,l}^3} \le 1 \tag{4.33}$$

$$-1 \le \frac{\left(g_1^i a_l \omega_{i,l}\right)^2 - \left(g_2^i a_l\right)^2 - \left(\omega_{i,l}\right)^4}{2 g_2^i a_l \omega_{i,l}^2} < 1. \tag{4.34}$$

Graphical Representation: Let

$$P_{i,l} = g_1^i a_l \omega_{i,l}\left[\sin \omega_{i,l} h_{i,1}^l \hat{\mathbf{e}}_1 + \cos \omega_{i,l} h_{i,1}^l \hat{\mathbf{e}}_2\right]$$

$$Q_{i,l} = g_2^i a_l\left[-\cos \omega_{i,l} h_{i,2}^l \hat{\mathbf{e}}_1 + \sin \omega_{i,l} h_{i,2}^l \hat{\mathbf{e}}_2\right]$$

$$R_{i,l} = (\omega_{i,l})^2 \hat{\mathbf{e}}_1,$$

then the first (X) and second (Y) components of the vector addition

$$P_{i,l} + Q_{i,l} = R_{i,l} \tag{4.35}$$

are equal to the first and second rows of Eq. (4.28), when $\theta_l = \pi$. In fact $g_1^i a_l \omega_{i,l}$, $g_2{}^i a_l$, and $(\omega_{i,l})^2$ are the side lengths of the delay characteristic triangle shown in Fig. 4.15. In order to assure existence of this characteristic triangle, $\omega_{i,l}$ must satisfy the following inequalities:

$$\begin{cases} g_1^i a_l \omega_{i,l} + g_2^i a_l > (\omega_{i,l})^2 \\ (\omega_{i,l})^2 + g_1^i a_l \omega_{i,l} > g_2^i a_l. \\ g_2^i a_l + (\omega_{i,l})^2 > g_1^i a_l \omega_{i,l} \end{cases} \tag{4.36}$$

Example 4.4. Consider an MAS that consists of $N = 8$ agents (3 leaders and 5 followers) and evolves in the $X - Y$ plane. Initial positions of the agents and corresponding parameters $\alpha_{i,1}$, $\alpha_{i,2}$, and $\alpha_{i,3}$ are listed in Table 4.3. Let followers apply the communication graph shown in Fig. 3.18 to acquire the desired positions defined by the homogeneous transformation through local communication, where communication weights are listed in Table 3.9. Note that the eigenvalues of the matrix A, $\lambda_1 = -a_1 = -0.1478$, $\lambda_2 = -a_2 = -0.6489$, $\lambda_3 = -a_3 = -1.0765$,

Fig. 4.15 Graphical representation of the delay characteristic triangle

$$O_1O_2 = g_1^i \omega_{i,l} a_l \qquad \alpha = \omega_{i,l} h_{i,1}$$
$$O_2O_3 = g_2^i a_l \qquad \beta = \omega_{i,l} h_{i,2}$$
$$O_3O_1 = (\omega_{i,l})^2$$

Table 4.3 Initial positions of the agents and corresponding parameters $\alpha_{i,1}, \alpha_{i,2}$, and $\alpha_{i,3}$ in Example 4.4

i	X_i	Y_i	$\alpha_{i,1}$	$\alpha_{i,2}$	$\alpha_{i,3}$
1	1.00	5.00	-	-	-
2	3.00	7.00	-	-	-
3	5.00	8.00	-	-	-
4	2.47	6.24	0.4943	0.2767	0.2290
5	3.24	6.97	0.1480	0.5857	0.2663
6	3.70	7.19	0.1577	0.3322	0.5101
7	2.85	6.55	0.3762	0.3214	0.3024
8	2.81	6.55	0.3539	0.3868	0.2593

Fig. 4.16 Paths chosen by the leaders in Example 4.4

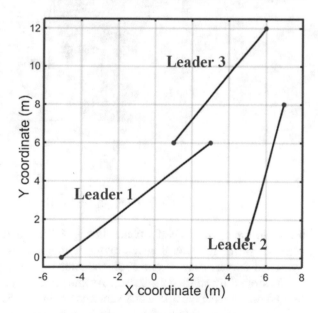

$\lambda_4 = -a_4 = -1.5116$, and $\lambda_5 = -a_5 = -1.6152$, are all real. Furthermore, leaders move on the straight lines shown in Fig. 4.16 with constant velocities.

MAS evolution without communication delay: Let $g_1 = g_1^i > 0$, $g_2 = g_2^i > 0$ and communication delays $h_{i,1}$ and $h_{i,2}$ be both zero ($\forall i \in V_F$). Then, MAS evolution dynamics is stable, because the roots of the characteristic equation

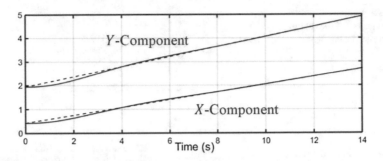

Fig. 4.17 X and Y components of the desired and actual positions of the follower 8

Fig. 4.18 $h_1{}^l - h_2{}^l$
($l = 1, 2, \ldots, 5$) curves
associated with different
eigenvalues of the matrix A;
Stability of MAS evolution is
gauranted, if $(h_{i,1}, h_{i,2})$ is
inside the shaded area
($\forall i \in V_F$)

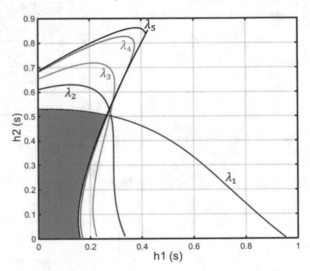

$$|s^2 I - (g_1 s + g_2)A| = 0$$

are all in the open left half s-plane. In Fig. 4.17, X and Y components of desired positions of the follower 8, given by the homogeneous transformation, are shown by dotted curves. Also, X and Y components of actual position of the follower 8 are illustrated by continuous curves. As shown, difference between the actual and desired positions of the follower 8 vanishes around $t = 5s$, while the follower 8 only accesses positions and velocities of the in-neighbor agents 4, 5, and 7.

MAS evolution with communication delay: Now suppose that the communication delays in Eq. (4.18) are not zero. It is assumed that $g_1 = g_1{}^i = 5$ and $g_2 = g_2{}^i = 10$, $h_1{}^l = h_{i,1}{}^l$ and $h_2{}^l = h_{i,2}{}^l$ ($i = 4, 5, \ldots, 8$ and $l = 1, 2, \ldots, 5$). $h_1{}^l - h_2{}^l$ associated with different eigenvalues of the matrix A are shown in Fig. 4.18.

In Fig. 4.19, components of desired and actual positions of the follower 8 are shown by dotted and continuous curves, respectively, where $h_{8,1} = 0.21s$ and

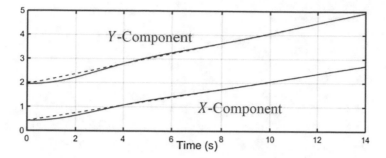

Fig. 4.19 X and Y components of the desired and actual positions of the follower 8 when $h_{8,1} = 0.21s$ and $h_{8,2} = 0.50s$

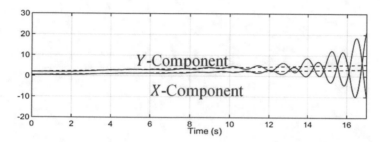

Fig. 4.20 X and Y components of desired and actual positions of the follower 8 when $h_{8,1} = 0.28s$ and $h_{8,2} = 0.50s$

$h_{8,2} = 0.50s$. Because $(h_{8,1}, h_{8,2}) = (0.21, 0.50)$ is inside the shaded stability region shown in Figure 6, follower 8 ultimately reaches the desired position given by a homogeneous transformation.

In Fig. 4.20, X and Y components of desired and actual positions of the follower 8 are shown, when $h_{8,2}$ is still 0.50 but $h_{8,1}$ is increased from $0.21s$ to $0.28s$. As it is observed evolution of the agent 8 is unstable because $(0.021, 0.50)$ is outside the shaded stability area.

Chapter 5
Alignment as Biological Inspiration for Control of Multi-Agent Systems

In this chapter, a framework for the evolution of an MAS under alignment strategy is developed. This novel idea comes from the hypothesis that natural biological swarms do not perform peer-to-peer communication to sustain the group behavior as a collective. The group evolution is more likely based on what each individual agent perceives of its nearby agent's behavior to control its own action. Most available engineering swarms rely on local interaction, where an individual agent requires precise state information of its neighboring agents to evolve. Here, agents of an MAS are considered as particles of a continuum (deformable Body) transforming under a homogeneous mapping. Using the key property of homogeneous transformation that two crossing straight lines in an initial configuration translate as two different crossing straight lines, agents can evolve collectively without peer-to-peer communication.

5.1 Initial Distribution of the Agents

Suppose that Ω_{t_0} is a closed domain in $\mathbb{R}^n$ enclosing a finite number of agents. Let N_l leader agents be placed at the boundary of the domain Ω_{t_0}, which is denoted by $\partial\Omega_{t_0}$. Leader agents occupy the opposite ends of m leading segments, where every interior (follower) agent is placed at the intersection of the two crossing line segments. The boundary (leader) agents are numbered $1, 2, \ldots, N_l$, and the follower agents are indexed $N_l + 1, N_l + 2, \ldots, N$. Shown in Fig. 5.1 is the domain Ω_{t_0} that is an example of a convex domain in $\mathbb{R}^2$. As illustrated, each follower agent is located at the intersection of the two leading line segment inside Ω_{t_0}.

Because the domain Ω_{t_0} is convex, every line contained in the interior of Ω_{t_0} crosses the boundary $\partial\Omega_{t_0}$ only at the two points.

© Springer International Publishing AG 2016
H. Rastgoftar, *Continuum Deformation of Multi-Agent Systems*,
DOI 10.1007/978-3-319-41594-9_5

Fig. 5.1 A convex domain Ω_{t_0} with some interior points resulting from intersection of some lines passing through Ω_{t_0} [112]

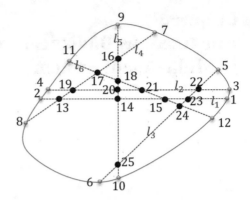

5.2   Evolution of Leader Agents

Let leaders' positions be given by a nonsingular mapping f at any time t, where f has the following properties:

- The Jacobian matrix $Q(R_i, t) = \dfrac{\partial f}{\partial R_i}$ has positive eigenvalues.
- $r_i(t_0) = f(R_i, t_0) = R_i$ (R_i is the initial position of the agent i.).

The following theorem proves that the convex region Ω_{t_0} at the initial time t_0 is transformed to another convex domain Ω_t, at time $t \geq t_0$, and the number of intersection points of the leading segments inside the regions Ω_{t_0} and Ω_t remains unchanged during transformation, if f is a nonsingular mapping and satisfies the above two properties.

Theorem 5.1. *Suppose that $\Omega_{t_0} \in \mathbb{R}^2$ is a convex region and four distinct points A, B, C, and D on Ω_{t_0} have the arrangement illustrated in Fig. 5.2. (If we move along the curve Ω_{t_0} in the counterclockwise direction, then A is met before B, B is met before C, C is met before D, and D is met before A.). If $\partial\Omega_{t_0}$ is mapped to another convex curve $\partial\Omega_t$, under a nonsingular mapping f (meeting properties 1 and 2), where the points A, B, C, and D, on the curve $\partial\Omega_{t_0}$, are mapped to A', B', C', and D', respectively, on the curve $\partial\Omega_t$, then*

- *Arrangement of the points A', B', C', and D' is the same as the arrangement of the points A, B, C, and D. (If we move along the curve $\partial\Omega_{t_0}$ in the counter clockwise direction, then A is met before B, B is met before C, C is met before D, and D is met before A.)*
- *Point i (positioned at the intersection of the line segments AC and BD) and point i' (positioned at the intersection of the line segments A'C' and B'D') are both placed inside the domains Ω_{t_0} and Ω_t, respectively.*

Proof. Every two different points on the closed curve $\partial\Omega_{t_0}$, at time t_0, is mapped to different points on the closed curve $\partial\Omega_t$, at the time $t \geq t_0$, if f satisfies properties 1 and 2. Therefore, the arrangement of the points A', B', C', and D', on the curve

Fig. 5.2 Schematic of a
homeomorphic mapping
between two convex curves
[112]

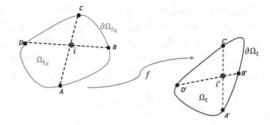

Fig. 5.3 Schematic of an
MAS transformation when
adjacency among the agents
is changed [112]

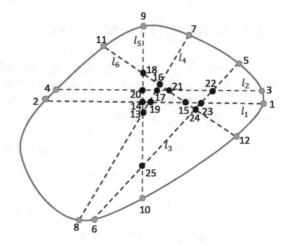

$\partial\Omega_t$ is the same as the ordering of A, B, C, and D on the curve $\partial\Omega_{t_0}$. By convexity
of the curve $\partial\Omega_{t_0}$ points on the line segments $A'C'$ and $B'D'$ does not leave Ω_t.
This implies that the intersection of the segments $A'C'$ and $B'D'$ is located inside the
domain Ω_t.

Theorem 5.1 allows to find requirements under which the number of intersection
points (occupied by followers) remains invariant. However, it is not still guaranteed
that adjacency among agents is preserved. For example, consider the configuration
of the MAS, shown in Fig. 5.1, that is transformed into the configuration Ω_t, shown
in Fig. 5.3, at time $t \geq t_0$. As it is seen, Ω_t is still convex and the arrangement of
the leaders and the total number of intersection points both remain unchanged. It
is noted that the configuration Ω_t (shown in Fig. 5.3) is a possible formation that
is resulted from a slight counterclockwise perturbation of the leader 8 along the
boundary $\partial\Omega_{t_0}$. However, adjacency among some agents changes considerably due
to this small perturbation. For instance, while the follower 17 is adjacent to the
followers 11, 16, 18, and 19 in Fig. 5.1, in Fig. 5.3, the followers 16, 19, 20, and 21
are adjacent to the follower 17.

Homogeneous Transformation under Alignment Strategy : If the leaders' positions are defined by a homogeneous deformation at any time t, the total number of agents and adjacency among the agents are both preserved. Throughout this chapter, it is assumed that leaders transform under a homogeneous mapping, therefore the initial and current positions of the leader i ($i = 1, 2, \ldots, N_l$) satisfy Eq. (2.2).

5.3   Evolution of Follower Agents

5.3.1   Followers' Desired Positions

It is assumed that follower agent i is adjacent with agents i_1, i_2, i_3, and i_4. It is supposed that $S_{i,1}$ is a leading line segment whose end points are occupied by i_1 and i_2 and $S_{i,2}$ is a line segment whose end points are occupied by i_3, and i_4. These line segments can be defined by the following parametric forms:

$$S_{1,i}: \begin{cases} x(t) = (x_{i_2}(t) - x_{i_1}(t))\, h_{i,1,2} + x_{i_1}(t) \\ y(t) = (y_{i_2}(t) - y_{i_1}(t))\, h_{i,1,2} + y_{i_1}(t) \end{cases} \tag{5.1}$$

$$S_{2,i}: \begin{cases} x(t) = (x_{i_4}(t) - x_{i_4}(t))\, h_{i,3,4} + x_{i_3}(t) \\ y(t) = (y_{i_4}(t) - y_{i_4}(t))\, h_{i,3,4} + y_{i_3}(t) \end{cases} . \tag{5.2}$$

where $0 \le h_{i,1,2} \le 1$ and $0 \le h_{i,3,4} \le 1$. Intersection point of two crossing segments $S_{i,1}$ and $S_{i,2}$ is considered as the desired position of the follower i at a time t is denoted by $r_{i,d} = x_{i,d}\hat{\mathbf{e}}_x + y_{i,d}\hat{\mathbf{e}}_y$. Components of $r_{i,d}$ is given by:

$$\begin{cases} x_{i,d}(t) = (x_{i_2}(t) - x_{i_1}(t))hd_{i,1,2} + x_{i_1}(t) \\ y_{i,d}(t) = (y_{i_2}(t) - y_{i_1}(t))hd_{i,1,2} + y_{i_1}(t) \end{cases} \tag{5.3}$$

$$\begin{cases} x_{i,d}(t) = (x_{i_4}(t) - x_{i_3}(t))hd_{i,3,4} + x_{i_3}(t) \\ y_{i,d}(t) = (y_{i_4}(t) - y_{i_3}(t))hd_{i,3,4} + y_{i_3}(t) \end{cases} . \tag{5.4}$$

Fig. 5.4 Schematic of evolution of follower i under alignment strategy [112]

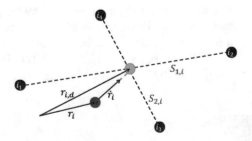

Notice that $hd_{i,1,2}$ and $hd_{i,3,4}$ are the unique parameters specifying positions of the intersection point. By equating the right-hand sides of Eqs. (5.3) and (5.4), it is concluded that

$$\begin{bmatrix} x_{i_2} - x_{i_1} & x_{i_3} - x_{i_4} \\ y_{i_2} - y_{i_1} & y_{i_3} - y_{i_4} \end{bmatrix} \begin{bmatrix} hd_{i,1,2} \\ hd_{i,3,4} \end{bmatrix} = \begin{bmatrix} x_{i_3} - x_{i_1} \\ y_{i_3} - y_{i_1} \end{bmatrix}. \tag{5.5}$$

Therefore, the X and Y components of the desired position of the follower i become

$$\begin{bmatrix} x_{i,d} \\ y_{i,d} \end{bmatrix} = \begin{bmatrix} x_{i_2} - x_{i_1} & 0 \\ 0 & y_{i_4} - x_{i_3} \end{bmatrix} \begin{bmatrix} x_{i_2} - x_{i_1} & x_{i_3} - x_{i_4} \\ y_{i_2} - y_{i_1} & y_{i_3} - y_{i_4} \end{bmatrix}^{-1} \begin{bmatrix} x_{i_3} - x_{i_1} \\ y_{i_3} - y_{i_1} \end{bmatrix} + \begin{bmatrix} x_{i_1} \\ y_{i_3} \end{bmatrix}. \tag{5.6}$$

In the theorem below, it is shown that the parameters $hd_{i,1,2}$ and $hd_{i,3,4}$ remain time invariant if the MAS transform under a homogeneous map.

Theorem 5.2 ([109, 112]). *Under a homogeneous deformation of the two crossing line segments $S_{1,i}$, $S_{2,i}$, the parameters $hd_{i,1,2}$ and $hd_{i,3,4}$ (corresponding to the intersection point) remain positive and time invariant. Hence, $hd_{i,1,2}$ and $hd_{i,3,4}$ are denoted by $HD_{i,1,2}$ and $HD_{i,3,4}$ and determined based on the initial positions of the agents i_1, i_2, i_3, and i_4 as follows:*

$$\begin{bmatrix} HD_{i,1,2} \\ HD_{i,3,4} \end{bmatrix} = \begin{bmatrix} X_{i_2} - X_{i_1} & X_{i_3} - X_{i_4} \\ Y_{i_2} - Y_{i_1} & Y_{i_3} - Y_{i_4} \end{bmatrix}^{-1} \begin{bmatrix} X_{i_3} - X_{i_1} \\ Y_{i_3} - Y_{i_1} \end{bmatrix}. \tag{5.7}$$

Proof. When agents i_1, i_2, i_3, and i_4 deform under a homogeneous transformation, position of the agent i_4 can be uniquely expanded as the linear combination of the positions of agents i_1, i_2, and i_3 as follows:

$$\begin{bmatrix} x_{i_4} \\ y_{i_4} \end{bmatrix} = \alpha_{i_4,i_1} \begin{bmatrix} x_{i_1} \\ y_{i_1} \end{bmatrix} + \alpha_{i_4,i_2} \begin{bmatrix} x_{i_2} \\ y_{i_2} \end{bmatrix} + \alpha_{i_4,i_3} \begin{bmatrix} x_{i_3} \\ y_{i_3} \end{bmatrix} \tag{5.8}$$

where the parameters α_{i_4,i_1}, α_{i_4,i_2}, and α_{i_4,i_3} remain unchanged (See Section 2.1.1 and Eqs. (2.14) and (2.15).) By substituting x_{i_4} and y_{i_4} in Eq. (5.4) by Eq. (5.8), and then equating the right-hand sides of Eqs. (5.3) and (5.4), it is concluded that

$$\begin{bmatrix} x_{i_2} - x_{i_1} & \alpha_{i_4,i_1}(x_{i_3} - x_{i_1}) + \alpha_{i_4,i_2}(x_{i_3} - x_{i_2}) \\ y_{i_2} - y_{i_1} & \alpha_{i_4,i_1}(y_{i_3} - y_{i_1}) + \alpha_{i_4,i_2}(y_{i_3} - y_{i_2}) \end{bmatrix} \begin{bmatrix} HD_{i,1,2} \\ HD_{i,3,4} \end{bmatrix} = \begin{bmatrix} x_{i_3} - x_{i_1} \\ y_{i_3} - y_{i_1} \end{bmatrix}. \tag{5.9}$$

Equation (5.9) can be rewritten as follows:

$$(1 - HD_{i,1,2} - \alpha_{i_4,i_1} HD_{i,3,4}) x_{i_1} + (HD_{i,1,2} - \alpha_{i_4,i_2} HD_{i,3,4}) x_{(i_2)} +$$
$$(HD_{i,3,4}(\alpha_{i_4,i_1} + \alpha_{i_4,i_2}) - 1) x_{i_3} = 0 \tag{5.10}$$

$$(1 - HD_{i,1,2} - \alpha_{i_4,i_1} HD_{i,3,4})y_{i_1} + (HD_{i,1,2} - \alpha_{i_4,i_2} HD_{i,3,4})y(i_2) +$$

$$(HD_{i,3,4}(\alpha_{i_4,i_1} + \alpha_{i_4,i_2}) - 1)y_{i_3} = 0. \tag{5.11}$$

By considering Eqs. (5.10) and (5.11), it is concluded that

$$\begin{bmatrix} 1 & -\alpha_{i_4,i_2} \\ 1 & \alpha_{i_4,i_1} \end{bmatrix} \begin{bmatrix} HD_{i,1,2} \\ HD_{i,3,4} \end{bmatrix} = \begin{bmatrix} 0 \\ 1 \end{bmatrix}. \tag{5.12}$$

Hence,

$$HD_{i,1,2} = \frac{\alpha_{i_4,i_2}}{\alpha_{i_4,i_1} + \alpha_{i_4,i_2}} \tag{5.13}$$

$$HD_{i,3,4} = \frac{1}{\alpha_{i_4,i_1} + \alpha_{i_4,i_2}} \tag{5.14}$$

remain time invariant, if agents i_1, i_2, i_3, and i_4 deform under a homogeneous mapping. Observe that the plane of motion can be divided into seven subregions based on the signs of the parameters α_{i_4,i_1}, α_{i_4,i_2}, and α_{i_4,i_3} as illustrated in Fig. 5.5. As observed, α_{i_4,i_1} and α_{i_4,i_2} are both positive, making $HD_{i,1,2}$ and $HD_{i,3,4}$ positive as well. Therefore, $HD_{i,1,2}$ and $HD_{i,3,4}$ are obtained from the initial positions of the agents i_1, i_2, i_3, and i_4 by using Eq. (5.7).

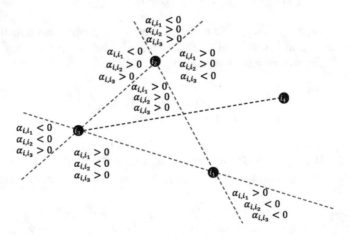

Fig. 5.5 Dividing motion plane to seven subregions based on the signs of the parameters α_{i_4,i_1}, α_{i_4,i_2}, and α_{i_4,i_3} [112]

5.3.2   Perception Weights

By considering Eqs. (5.3) and (5.4), the desired position $r_{i,d}$ can be expressed as

$$r_{i,d} = w_{P_{i,i_1}} r_{i_1} + w_{P_{i,i_2}} r_{i_2} + w_{P_{i,i_3}} r_{i_3} + w_{P_{i,i_4}} r_{i_4}, \tag{5.15}$$

where

$$\begin{cases} w_{P_{i,i_1}} = \dfrac{1}{2}(1 - hd_{i,1,2}) \\ w_{P_{i,i_2}} = \dfrac{1}{2} hd_{i,1,2} \\ w_{P_{i,i_3}} = \dfrac{1}{2}(1 - hd_{i,3,4}) \\ w_{P_{i,i_2}} = \dfrac{1}{2} hd_{i,3,4} \end{cases} \tag{5.16}$$

are called perception weights. Since $0 < hd_{i,1,2} < 1$ and $0 < hd_{i,1,2} < 1$ are obtained by solving Eq. (5.5), perception weights remain positive during evolution, where

$$w_{P_{i,i_1}} + w_{P_{i,i_2}} + w_{P_{i,i_3}} + w_{P_{i,i_4}} = 1. \tag{5.17}$$

Now, we can define a time varying perception weight matrix $W_p \in \mathbb{R}^{(N-N_l) \times N}$ with the ij entry that is defined as follows:

$$W_{Pij} = \begin{cases} w_{P_{i+N_l,j}}(t) > 0 & \text{if } j \in N_{i+N_l} \\ -1 & \text{if } i + N_l = j \,. \\ 0 & \text{Otherwise} \end{cases} \tag{5.18}$$

Notice that the N_{i+N_l} defines indices of the in-neighbor agents of the follower $i + N_l \in V_F$. Also, W_p is zero-sum-row (sum of each row of the matrix W_P is zero.). Let W_P be partitioned as follows:

$$W_P = \begin{bmatrix} B_p & A_P \end{bmatrix} \in \mathbb{R}^{(N-N_l) \times N}. \tag{5.19}$$

Then the matrix $A_P(t) \in \mathbb{R}^{(N-N_l) \times (N-N_l)}$ is Hurwitz. The proof is given in the Theorem 5.3.

5.3.3   Followers' Dynamics

Let each follower i update its current position according to

$$\dot{r}_i = g(r_{i,d} - r_i), \tag{5.20}$$

where $g \in \mathbb{R}_+$ is constant. Then the qth components of the followers' positions are updated by the following first order dynamics:

$$\dot{z}_q = g(A_p z_q + B_q u_q). \tag{5.21}$$

Notice that $u_q = [x_{q,1} \ldots x_{q,N_l}]^T \in \mathbb{R}^{N_l}$ and $z_q = [x_{q,N_l+1} \ldots x_{q,N}]^T \in \mathbb{R}^{N-N_l}$ denote the qth components of positions of the leaders and followers, respectively. It is noted that

$$A_p = -(I - F_p(t)) \tag{5.22}$$

is diagonally dominant, where $F_p(t)$ is a nonnegative and irreducible matrix with the spectral radius $\rho(F_p(t))$ that remains less than 1 during MAS evolution.

Theorem 5.3. *Consider the diagonally dominant time-varying matrix $A_p(t)$, where the sum of N_l ($1 \leq N_l < N$) rows of $A_P(t)$ is negative, the remaining rows of $A_P(t)$ are zero-sum, and $F_P(t)$ is an irreducible nonnegative time varying matrix. Then the linear time varying dynamics*

$$\dot{z}_q = A_p z_q \tag{5.23}$$

is asymptotically stable.

Proof. Because the sum of N_l rows of $A_P(t)$ is negative and $F_P(t)$ is an irreducible and nonnegative time-varying matrix, then the spectrum $\rho(F_P(t)) < 1$. Therefore, eigenvalues of $A_P(t)$ have the negative real part at any time t. Let

$$V = \frac{1}{2} z_q{}^T z_q \tag{5.24}$$

be considered as the Lyapunov function, then

$$\dot{V} = \dot{z}_q^T z_q = z_q{}^T A_p(t) z_q = -z_q{}^T (I - F_p(t)) z_q \leq -z_q{}^T (1 - \rho(F_P(t))) z_q < 0. \tag{5.25}$$

This is because $A_p(t)$ is diagonally dominant with the diagonal elements that are all -1. Consequently, the zero dynamics 5.23 is asymptotically stable.

Remark 5.1. Because $A_p(t)$ is Hurwitz at any time t, followers' transient positions asymptotically converge to the equilibrium state of the dynamics (5.21) that is given by

$$z_{q,s} = -A_p{}^{-1}(t_f) B_p(t_f) u_q(t_f) \tag{5.26}$$

where t_f is the final time when leaders stop. Because final configuration of the leaders at the time t_f is a homogeneous transformation of the leaders' initial

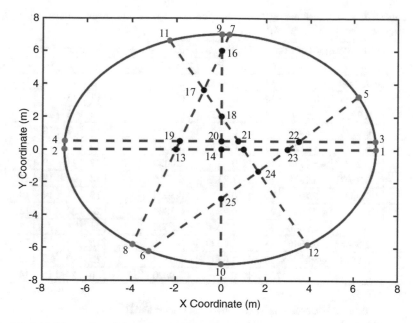

Fig. 5.6 Initial formation of the agents in Example 5.1

positions, $hd_{i,1,2}(t_f) = HD_{i,1,2}$, and $hd_{i,3,4}(t_f) = HD_{i,3,4}$ (See Theorem 5.2.), final configuration of the MAS is a homogeneous transformation of the agents' initial formation.

Example 4.1. Consider an MAS consisting of 25 agents moving in a plane. Shown in Fig. 5.6 is the initial distribution of the MAS, where leaders are all placed along the circle of radius $7m$ that is centered at the origin. Leaders are numbered by 1, 2, ..., 12, and they are placed at the end points of six leading line segments. Followers are indexed by 13, 14, ..., 25, where they are initially located inside the circle at intersection points of the leading segments.

In Table 5.1, initial positions of the leaders and followers as well as adjacency among the agents, are defined.

Evolution of the Leaders: Let the leaders evolve under the following homogeneous map

$$\begin{bmatrix} x_i(t) \\ y_i(t) \end{bmatrix} = \begin{bmatrix} 1+0.8t-0.04t^2 & 0.4t-0.02t^2 \\ 0.2t-0.01t^2 & 1+0.6t-0.03t^2 \end{bmatrix} \begin{bmatrix} X_i \\ Y_i \end{bmatrix} + \begin{bmatrix} -t+0.05t^2 \\ 0.6t-0.03t^2 \end{bmatrix} \quad 0 \le t \le 10$$

(5.27)

with the leaders coming to rest at $t = 10s$. Final positions of leader agents, as well as followers' final positions, are listed in Table 5.2. Final positions of followers also satisfy Eq. (5.27). This implies that ultimate formation of all agents is a homogeneous transformation of the initial configuration.

Table 5.1 Initial positions of leader and follower agents and index numbers of in-neighbor agents in Example 5.1

	Initial Positions of Leaders			Initial Positions of Followers		Adjacent Agents			
i	$X_i(m)$	$Y_i(m)$	i	$X_i(m)$	$Y_i(m)$	i_1	i_2	i_3	i_4
1	7.0000	0	13	−2.0000	0	2	14	8	19
2	−7.0000	0	14	0	0	13	15	25	20
3	6.9821	0.5000	15	1.0000	0	14	23	21	24
4	−6.9821	0.5000	16	0	6.0000	17	7	18	9
5	6.2170	3.2170	17	−0.8000	3.6000	19	16	18	11
6	−3.2170	−6.2170	18	0	2.0000	17	21	20	16
7	0.3307	6.9921	19	−1.8333	0.5000	4	20	13	17
8	−3.9307	−5.7921	20	0	0.5000	19	21	14	18
9	0	7.0000	21	0.7500	0.5000	20	22	15	18
10	0	−7.0000	22	3.5000	0.5000	21	3	23	5
11	−2.3048	6.6096	23	3.0000	0	15	1	24	22
12	3.9048	−5.8096	24	1.6667	−1.3333	15	12	25	23
			25	0	−3.0000	6	24	10	14

Table 5.2 Desired final positions of the leaders and followers in Example 5.1

	Initial Positions of Leaders			Initial Positions of Followers	
i	$X_i(m)$	$Y_i(m)$	i	$X_i(m)$	$Y_i(m)$
1	30.0000	10.0000	13	−15.0000	1.0000
2	−40.0000	−4.0000	14	−4.9999	3.0000
3	30.9105	11.9821	15	0.0000	4.0000
4	−38.9105	−1.9821	16	7.0000	27.0000
5	32.5190	22.0850	17	−1.8000	16.6000
6	−33.5190	−25.0850	18	−1.0000	10.9999
7	10.6377	31.2991	19	−13.1667	3.1667
8	−36.2377	−24.0991	20	−3.9999	5.0000
9	9.0000	31.0000	21	−0.2500	5.7500
10	−19.0000	−25.0000	22	13.5000	8.5000
11	−3.3048	27.1336	23	10.0000	6.0000
12	2.9048	−16.3336	24	0.6667	−0.6667
			25	−10.9999	−8.9999

Evolution of the Followers: Each follower i ($i = 13, 14, \ldots, 15$) updates its position according to the dynamics (5.20), where $g = 10$. Note that the perception weights are obtained by using Eq. (5.16). In Fig. 5.7, perception weights of the follower 20 ($w_{P20,14}, w_{P20,18}, w_{P20,19}$, and $w_{P20,21}$) are shown versus time. As observed, both initial and final values of perception weights are the same. This implies that final formation of the MAS is a homogeneous transformation of the initial configuration.

Shown in Fig. 5.8 are the paths of the agents 16, 17, 18, 20, and 21 in the $X - Y$ plane. Also, positions of the agents at $t = 5s$ are shown in Fig. 5.8. It is seen that the

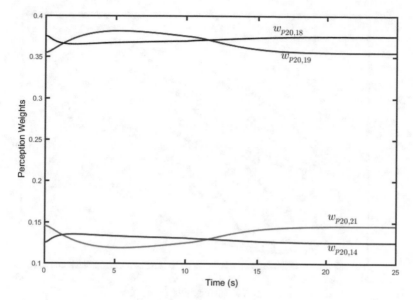

Fig. 5.7   Perception weights of the follower 23

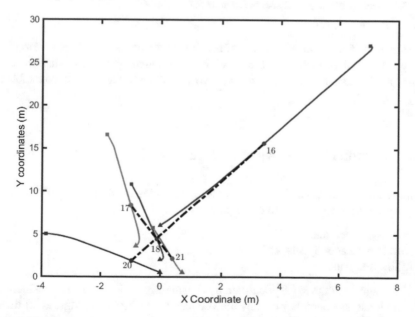

Fig. 5.8   Paths of the agents 16, 17, 18, 20, and 21 in the $X - Y$; configuration of the agents 16, 17, 18, 20, and 21 at $t = 5s$

follower 18 is located at the intersection of the two crossing line segments (follower 18 is aligned with the agents 16 and 20, as well as the agents 17 and 21). Note that the initial and final positions of agents are shown by $\triangle$ and $\square$, respectively.

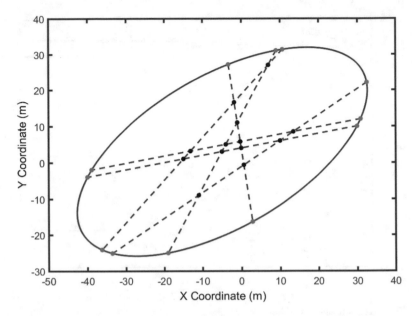

Fig. 5.9 Final formation of the MAS in Example 5.1

Shown in Fig. 5.9 are the final positions of 25 agents of the MAS, in the $X - Y$ plane. The final formation of the MAS is a homogeneous transformation of the initial configuration. Final positions of the follower agents are also listed in Table 5.2 (Fig. 5.9).

5.4   Alignment Using Agents' Triangulation

In this section, leaders are distributed along the sides of the leading triangle, and agents are categorized into three different groups:

- primary leader gents,
- secondary leader agents, and
- follower agents.

Three agents of the MAS, identified by the numbers 1, 2, and 3, are considered as primary leaders, where they occupy the vertices of the leading triangle at all times t during MAS evolution. Notice that primary leaders evolve independently. It is noted that entries of the Jacobian Q and the vector D of a homogeneous deformation are uniquely determined based on the trajectories chosen by the primary leaders (See Eq. (2.21).)

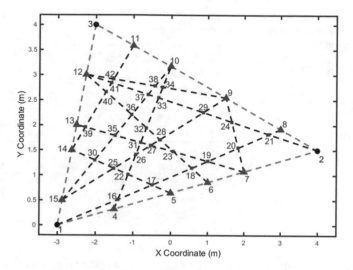

Fig. 5.10   Initial formation of the MAS in Example 5.2 [112]

Agents located on the sides of the leading triangle are considered as secondary leaders. Secondary leaders are at the end points of the leading line segments. They can acquire the desired positions, given by a homogeneous map and prescribed by the primary leaders, either through no communication or local interaction.

Followers are initially located at the intersection points of the crossing leading segments. Follower agents continuously attempt to preserve alignment with four neighboring agents.

A typical initial distribution for an MAS evolving based on agent' triangulation approach is shown in Fig. 5.10. As illustrated in Fig. 5.10, primary leaders are identified by 1, 2, and 3, secondary leaders are indexed by the numbers 4, 5, ..., 15, and followers are numbered 16, 17, ..., 32.

5.4.1   Evolution of the Secondary Leaders

In this section it is described how secondary leaders can acquire the desired positions either via no communication or local interaction.

5.4.1.1   Evolution of the Secondary Leaders under No Communication

Let position of the secondary leader i be uniquely expressed as the linear combination of the positions of the primary leaders 1, 2, and 3 according to Eq. (2.17), where parameters $\alpha_{i,1}$, $\alpha_{i,2}$, and $\alpha_{i,3}$ are uniquely determined by using Eq. (2.18), based on the initial positions of the secondary leader i and the primary leaders 1, 2, and 3. Secondary leaders can acquire desired positions given by a homogeneous

Table 5.3 Initial positions of the secondary leaders shown in Fig. 4.10, and the corresponding parameters $\alpha_{i,1}$, $\alpha_{i,2}$, and $\alpha_{i,3}$ [112]

i	$X_i(m)$	$Y_i(m)$	$\alpha_{i,1}$	$\alpha_{i,2}$	$\alpha_{i,3}$
4	−1.5000	0.3214	0.7857	0.2143	0
5	0	0.6429	0.5714	0.4286	0
6	1.0000	0.8571	0.4286	0.5714	0
7	2.0000	1.0714	0.2857	0.7143	0
8	3.0000	1.9167	0	0.8333	0.1667
9	1.5000	2.5417	0	0.5833	0.4167
10	0	3.1667	0	0.3333	0.6667
11	−1.0000	3.5833	0	0.1667	0.8333
12	−2.2500	3.0000	0.2500	0	0.7500
13	−2.5000	2.0000	0.5000	0	0.5000
14	−2.6250	1.5000	0.6250	0	0.3750
15	−2.8750	0.5000	0.8750	0	0.1250

deformation without local communication, only by knowing positions of the primary leaders in a finite horizon of time $t \in [0, T]$. Initial positions of the secondary leaders, shown in Fig. 5.10, and the corresponding parameters $\alpha_{i,k}$ ($i = 4, \ldots, 15$ and $k = 1, 2, 3$) are listed in Table 5.3.

5.4.1.2   Evolution of the Secondary Leaders under Local Communication

Let each secondary leader update its current position based on positions of the two adjacent leaders, where communication weights are determined based on the agents' positions in the initial configuration. Assume that the secondary leader i, with the initial position R_i, accesses positions of the adjacent secondary leaders, i_1 and i_2, with the initial positions R_{i_1}, and R_{i_2}. Then the communication weights are defined as follows:

$$w_{i,i_1} = \frac{\|R_i - R_{i_2}\|}{\|R_{i_1} - R_{i_2}\|} \tag{5.28}$$

$$w_{i,i_2} = \frac{\|R_i - R_{i_1}\|}{\|R_{i_1} - R_{i_2}\|} \tag{5.29}$$

Because the secondary leader i and the leaders, i_1 and i_2 are initially aligned,

$$w_{i,i_1} + w_{i,i_2} = 1. \tag{5.30}$$

Table 5.4 Communication weights of the secondary leaders for the configuration shown in Fig. 5.10; Secondary leaders' index numbers; Adjacent agents' communication weights [112]

Secondary Leaders' Index Numbers	Adjacent Agents		Communication Weights	
	i_1	i_2	w_{i,i_1}	w_{i,i_2}
4	1	5	0.5000	0.5000
5	4	6	0.4000	0.6000
6	5	7	0.5000	0.5000
7	6	2	0.6667	0.3333
8	2	9	0.6000	0.4000
9	8	10	0.5000	0.5000
10	9	11	0.4000	0.6000
11	10	3	0.5000	0.5000
12	3	13	0.5000	0.5000
13	12	14	0.3333	0.6667
14	13	15	0.6667	0.3333
15	14	1	0.3333	0.6667

Communication weights of the secondary leaders, that are consistent with the initial formation shown in Fig. 5.10, are listed in Table 5.4.

Weight Matrix: The ij entry of the matrix $W_c \in \mathbb{R}^{(N_l-3)\times N_l}$ is defined by

$$W_{cij} = \begin{cases} w_{i+3,j} > 0 & \text{if } j \in N_{i+3} \\ -1 & \text{if } j = i+3 \ . \\ 0 & \text{Otherwise} \end{cases} \qquad (5.31)$$

W_c is partitioned as follows:

$$W_c = \begin{bmatrix} B_c & A_c \end{bmatrix}. \qquad (5.32)$$

The matrix $A_c \in \mathbb{R}^{(N_l-3)\times(N_l-3)}$ is diagonally dominant and Hurwitz. For the initial positions of the secondary leaders, shown in Fig. 5.10, the partitions A_c and B_c of the matrix W_c become

$$
A_c =
\begin{bmatrix}
-1 & 0.75 & 0.25 & 0 & 0 & 0 & 0 & 0 & 0 & 0 & 0 & 0 \\
\dfrac{1}{3} & -1 & 0 & 0 & 0 & 0 & 0 & 0 & 0 & 0 & 0 & 0 \\
0.29 & 0 & -1 & 0.71 & 0 & 0 & 0 & 0 & 0 & 0 & 0 & 0 \\
0 & 0 & 0.64 & -1 & 0 & 0 & 0 & 0 & 0 & 0 & 0 & 0 \\
0 & 0 & 0 & 0 & -1 & 0.4 & 0 & 0 & 0 & 0 & 0 & 0 \\
0 & 0 & 0 & 0 & 0.5 & -1 & 0.5 & 0 & 0 & 0 & 0 & 0 \\
0 & 0 & 0 & 0 & 0 & \dfrac{2}{3} & -1 & \dfrac{1}{3} & 0 & 0 & 0 & 0 \\
0 & 0 & 0 & 0 & 0 & 0 & 0.25 & -1 & 0 & 0 & 0 & 0 \\
0 & 0 & 0 & 0 & 0 & 0 & 0 & 0 & -1 & 0.5 & 0 & 0 \\
0 & 0 & 0 & 0 & 0 & 0 & 0 & 0 & 0.625 & -1 & 0.375 & 0 \\
0 & 0 & 0 & 0 & 0 & 0 & 0 & 0 & 0 & 0.375 & -1 & 0.625 \\
0 & 0 & 0 & 0 & 0 & 0 & 0 & 0 & 0 & 0 & 0.4 & -1
\end{bmatrix}
\tag{5.33}
$$

$$
B_c{}^T =
\begin{bmatrix}
0 & 2/3 & 0 & 0 & 0 & 0 & 0 & 0 & 0 & 0 & 0 & 0.6 \\
0 & 0 & 0.36 & 0.6 & 0 & 0 & 0 & 0 & 0 & 0 & 0 & 0 \\
0 & 0 & 0 & 0 & 0 & 0 & 0.75 & 0.5 & 0 & 0 & 0 & 0
\end{bmatrix}.
\tag{5.34}
$$

Dynamics of the Secondary Leaders: Let each secondary leader i update its current position according to the first order dynamics (5.20), where

$$
r_{i,d} = \sum_{j \in N_i} w_{i,j} r_j.
\tag{5.35}
$$

Then the qth components of the positions of the secondary leaders are updated as follows:

$$
\dot{z}_{sl,q} = g \left(A_c z_{sl,q} + B_c u_{p,q} \right)
\tag{5.36}
$$

where

$$
u_{p,q} = \left[x_{q,1}(t)\ x_{q,2}(t)\ x_{q,3}(t) \right]^T
$$

$$
z_{sl,q} = \left[x_{q,4}(t) \ldots x_{q,N_l}(t) \right]^T
$$

denote the qth components of the positions of the primary and secondary leaders, respectively. As aforementioned, A_c is Hurwitz because all communication weights are positive. Therefore, the qth components of the positions of the secondary leaders asymptotically converge to the equilibrium state that is obtained as follows:

$$
z_{f,sl,q} = -A_c^{-1} B_c u_{p,q}(t_f).
\tag{5.37}
$$

Notice that t_f is the time when leaders stop and

$$Z_{f,sl,q} = \left[X_{f,q,4} \; \cdots \; X_{f,q,N_l}\right]^T$$

denotes the qth components of the final positions of the secondary leaders. Final configuration of the secondary leaders is a homogeneous transformation of the initial formation of the secondary leaders.

Remark 5.2. Although final formation of the secondary leaders is a homogeneous transformation of the initial configuration, they deviate from the desired positions prescribed by a homogeneous deformation during transition. However, an upper limit δ can be specified for the deviation of the secondary leaders from the desired positions by using Eq. (3.45) (See Section 3.2.3.1). This upper limit depends on (i) maximum of the magnitude of velocities of all primary leaders, (ii) the control gain g applied by the secondary leaders, (iii) total number of the (primary and secondary) leaders, (iv) dimension of motion space ($n = 2$), and (iv) $||A_c^{-1}||$.

5.4.2   Evolution of the Followers

Follower agents update their positions according to Eq. (5.20), where X and Y components of $r_{i,d}$ are determined by Eq. (5.6). Also, the invariant parameters $HD_{i,1,2}$ and $HD_{i,3,4}$ are uniquely determined by using Eq. (5.12) and listed in Table 5.5.

Remark 5.3 (Evolution of Followers under Local Communication). Evolution of an MAS under alignment strategy is advantageous because followers only need the direction information of the adjacent agents to learn the desired positions. They do not need to do peer to peer communication requiring exact positions of the adjacent agents. The alignment strategy can be employed by the followers to acquire a desired homogeneous transformation through local communication. Given positions of in-neighbor agents i_1, i_2, i_3, and i_4, follower i determines $HD_{i,1,2}$ and $HD_{i,3,4}$ at the initial time t_0 and then updates its position by

$$\dot{r}_i(t) = g\left(w_{Fi,i_1} r_{i_1} + w_{Fi,i_2} r_{i_2} + w_{Fi,i_3} r_{i_3} + w_{Fi,i_4} r_{i_4} - r_i(t)\right) \qquad (5.38)$$

where

$$\begin{cases} w_{Fi,i_1} = \dfrac{1}{2}\left(1 - HD_{i,1,2}\right) \\[2mm] w_{Fi,i_2} = \dfrac{1}{2} HD_{i,1,2} \\[2mm] w_{Fi,i_3} = \dfrac{1}{2}\left(1 - HD_{i,3,4}\right) \\[2mm] w_{Fi,i_4} = \dfrac{1}{2} HD_{i,3,4} \end{cases} \qquad (5.39)$$

Table 5.5 Time-invariant parameters $HD_{i,1,2}$ and $HD_{i,3,4}$ and the followers' initial perception weights in Example 5.2 [112]

i	i_1	i_2	i_3	i_4	$HD_{i,1,2}$	$HD_{i,3,4}$	w_{i,i_1}	w_{i,i_2}	w_{i,i_4}	w_{i,i_4}
16	1	17	4	22	0.6370	0.2741	0.1815	0.3185	0.3630	0.1370
17	16	18	5	22	0.4627	0.4302	0.2687	0.2313	0.2849	0.2151
18	17	19	6	23	0.7113	0.4628	0.1444	0.3556	0.2686	0.2314
19	18	20	7	23	0.3388	0.5185	0.3306	0.1694	0.2407	0.2593
20	19	21	7	24	0.4980	0.4794	0.2510	0.2490	0.2603	0.2397
21	20	8	2	24	0.7281	0.5632	0.1360	0.3640	0.2184	0.2816
22	16	26	17	25	0.5593	0.6322	0.2203	0.2797	0.1839	0.3161
23	18	28	19	27	0.5807	0.6064	0.2097	0.2903	0.1968	0.3032
24	20	9	21	29	0.5171	0.5641	0.2414	0.2586	0.2179	0.2821
25	15	26	22	30	0.6998	0.4288	0.1501	0.3499	0.2856	0.2144
26	25	27	22	31	0.5968	0.6226	0.2016	0.2984	0.1887	0.3113
27	26	28	23	31	0.6198	0.6916	0.1901	0.3099	0.1542	0.3458
28	27	29	23	32	0.1724	0.5024	0.4138	0.0862	0.2488	0.2512
29	28	9	24	33	0.6506	0.3954	0.1747	0.3253	0.3023	0.1977
30	15	35	25	14	0.6200	0.4500	0.1900	0.3100	0.2750	0.2250
31	26	32	27	35	0.4502	0.2821	0.2749	0.2251	0.3589	0.1411
32	31	33	28	36	0.3291	0.4750	0.3355	0.1645	0.2625	0.2375
33	32	34	29	37	0.7377	0.8145	0.1312	0.3688	0.0927	0.4073
34	33	10	9	38	0.3385	0.8948	0.3307	0.1693	0.0526	0.4474
35	30	36	31	39	0.5435	0.4647	0.2282	0.2718	0.2677	0.2323
36	35	37	32	40	0.5053	0.3653	0.2473	0.2527	0.3173	0.1827
37	36	38	33	41	0.7020	0.2197	0.1490	0.3510	0.3901	0.1099
38	37	10	34	42	0.3049	0.1547	0.3476	0.1524	0.4226	0.0774
39	14	40	35	13	0.3925	0.7754	0.3037	0.1963	0.1123	0.3877
40	39	41	36	12	0.7908	0.5662	0.1046	0.3954	0.2169	0.2831
41	40	42	37	12	0.7203	0.5929	0.1398	0.3602	0.2036	0.2964
42	41	11	38	12	0.0968	0.6032	0.4516	0.0484	0.1984	0.3016

are considered as the communication weights. Then, the qth components of the followers' positions are updated by

$$\dot{z}_q = g(A_P(t_0)z_q + B_P(t_0)u_q). \tag{5.40}$$

It is noticed that followers need to know the positions of the adjacent agents at any time t when they update their positions according to Eq. (5.38), although the communication weights are all time-invariant. In Table 5.5 followers' communication weights, that are consistent with the formation shown in Fig. 5.10, are calculated by using Eq. (5.39).

Example 5.3. Consider the MAS consisting of 42 agents (3 primary leaders, 12 secondary leaders, and 27 followers) with the initial configuration shown in Fig. 5.10. Let the primary leaders 1, 2, and 3 choose the paths shown in Fig. 5.11,

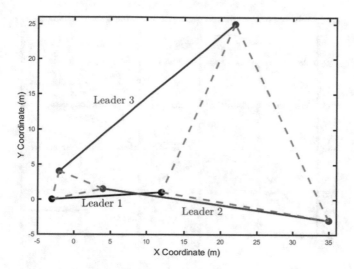

Fig. 5.11 Paths of the primary leaders [112]

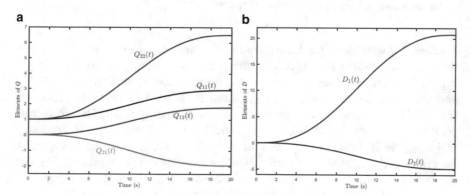

Fig. 5.12 Elements of Q and D versus time [112]

where they ultimately stop at $(X_{1,F}, Y_{1,F}) = (12, 1)$, $(X_{2,F}, Y_{2,F}) = (35, -3)$, and $(X_{3,F}, Y_{3,F}) = (22, 25)$, respectively, in $20s$. Entries of the Jacobian Q ($Q_{11}(t)$, $Q_{12}(t)$, $Q_{21}(t)$, and $Q_{22}(t)$) and vector D ($D_1(t)$ and $D_2(t)$), determined by the positions of the primary leaders by using Eq. (2.21), are shown in Fig. 5.12. As it is seen, $Q(0) = I \in \mathbb{R}^{2\times 2}$ and $D(0) = \mathbf{0} \in \mathbb{R}^2$.

Two scenarios are considered in this example. In the first scenario, secondary leaders acquire their desired positions by knowing positions of the primary leaders. In the second scenario, each secondary leader communicates with two adjacent agents to acquire the desired position given by a homogeneous deformation. Note that the desired homogeneous transformation is defined based on positions of the primary leaders.

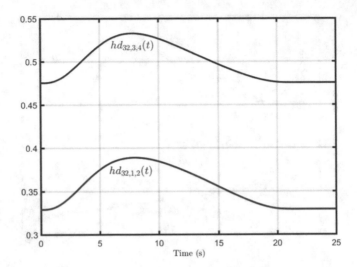

Fig. 5.13 Variations of $hd_{i,1,2}$ and $hd_{i,3,4}$ versus time for the follower 32, where secondary leaders evolve under no interagent communication [112]

Scenario I (Evolution of the Secondary Leaders under no Communication): Here the secondary leader i ($i = 4, 5, \ldots, 15$) updates its current position according to Eq. (2.17), where it knows positions of the primary leaders at all time $t \in [0, 20]s$ and the parameter $\alpha_{i,1}$, $\alpha_{i,2}$, and $\alpha_{i,3}$, listed in Table 5.3. Then the follower i ($i = 16, 17, \ldots, 42$) applies the proposed alignment strategy and updates its current position according to Eq. (5.20), where $g = 40$. In Fig. 5.13, the parameters of $hd_{i,1,2}$ and $hd_{i,3,4}$ are illustrated versus time for the follower 32. As it is seen in Fig. 5.13, $hd_{32,1,2}$ and $hd_{32,3,4}$ ultimately meet their corresponding initial values implying that the final position of follower 32 satisfies the condition of a homogeneous map.

In Fig. 5.14, configurations of the agents at four different times $t = 5s$, $t = 12s$, $t = 17s$, and $t = 20s$ are shown. As shown, followers ultimately reach the desired positions at the intersections of the leading line segments. This implies that the final formation of the MAS is a homogeneous transformation of the initial configuration.

Evolution of the secondary leaders under local communication: Let each secondary leader i interacts with two adjacent leaders and updates its position according to Eq. (5.20), where in-neighbor agents' index numbers as well as communication weights, obtained from, Eqs. (5.28) and (5.29), are as listed in Table 5.4. Here, each follower i updates its position according to the proposed alignment method similar to the scenario I. It is noted that both secondary leaders and followers apply the same control gain $g = 40$ to update their positions. Shown in Fig. 5.15 are the parameters $hd_{32,1,2}$ and $hd_{32,3,4}$ versus time for the follower 32. As illustrated, initial and final extents of $hd_{32,1,2}$ and $hd_{32,3,4}$ are the same. In Fig. 5.16, configurations of the agents at four sample times $t = 5s$, $t = 12s$, $t = 17s$, and $t = 20s$ are illustrated. It is observed that the final configuration of the MAS is the homogeneous transformation of the initial formation of the agents.

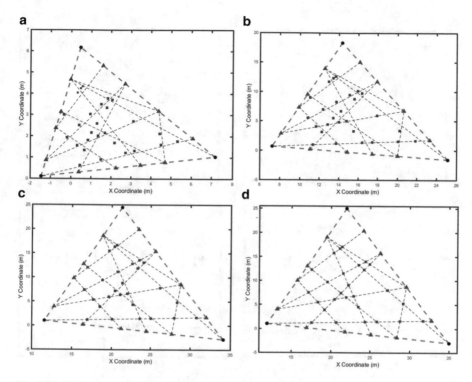

Fig. 5.14 Formations of the MAS in Example 5.2 at four different times $t = 5s$, $t = 12s$, $t = 17s$, and $t = 20s$, where secondary leaders evolve under no interagent communication [112]

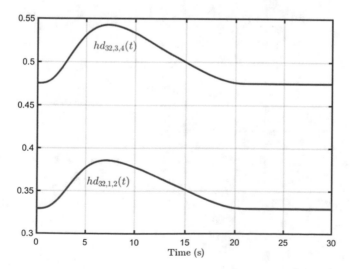

Fig. 5.15 Variations of $hd_{i,1,2}$ and $hd_{i,3,4}$ versus time for the follower 32, where secondary leaders update their positions through local interagent communication [112]

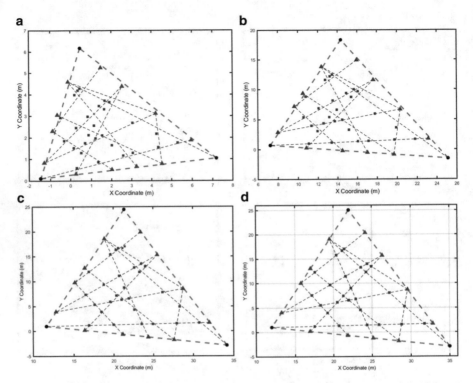

Fig. 5.16 Formations of the MAS in Example 5.2 at four different times (**a**) t=5s, (**b**) t=12s, (**c**) t=17s, and (**d**) t=20s, respectively, where secondary leaders update their positions through local interagent communication [112]

Chapter 6
Deployment of a Multi-Agent System on a Desired Formation

In this chapter, a decentralized control approach for deployment of an arbitrary distribution of a multi-agent system (MAS) on a desired formation in $\mathbb{R}^n$ ($n = 1, 2, 3$) is proposed, where avoidance of interagent collision is addressed. For this purpose, the motion of an MAS in $\mathbb{R}^n$ is decoupled into n separate 1-D motion problems. For evolution of the qth ($q = 1, 2, 3$) components of the agents' positions, two q-leaders are considered, where they guide the qth components of the MAS evolution. The remaining agents are considered as q-followers, where they update the qth components of their positions through local communication with the communication weights that are consistent with the qth components of the agents' positions in the final configuration.

6.1 Definitions

We use the following terms and definitions in this chapter:

Definition 6.1. An agent is called a *q-agent* when evolution of the qth component of its position is considered.

Definition 6.2. A q-agent i is called a *q-leader*, if the qth component of its position evolves independently. Notice that q-leaders are placed at the boundary of the desired formation.

Definition 6.3. An agent i is called a *q-follower*, if the qth component of its state (position and velocity) is updated based on the qth components of the positions of the two in-neighbor agents.

© Springer International Publishing AG 2016
H. Rastgoftar, *Continuum Deformation of Multi-Agent Systems*,
DOI 10.1007/978-3-319-41594-9_6

Definition 6.4. Any agent is identified by a number which is called *index number*. Notice that the set

$$V = \{1, 2, \ldots, N\}$$

defines index numbers of the agents. The index number of the agent $i \in V$ is called *q-index number*, when the qth component of position of the agent $i \in V$ is considered.

Definition 6.5. The communication graph, used by q-followers for updating the qth components of their positions, is called *q-path*. A q-path consists of N vertices, and it is obtained by arranging agents on a straight line from left to right based on the magnitude of the qth components of their positions in the desired final configuration. (The qth components of the q-agents' positions are increasing from left to right.). For this purpose, the furthest left and the furthest right agents are considered as the q-leaders, and the remaining agents are considered as the q-followers. For instance consider the desired configuration illustrated in Fig. 6.1 and Fig. 6.2. In Table 6.1 positions of the agents, shown in Fig. 6.1, and the corresponding q-index numbers of the agents are listed. Then, 1-path and 2-path graphs corresponding to the first (X) and second (Y) components of the agents' positions are illustrated in Fig. 6.2. As seen, 1-agents 1 and 11 are the two 1-leaders, and the remaining 1-agents are 1-followers. Furthermore, 2-agents 5 and 7 are the two 2-leaders, and the remaining 2-agents are 2-followers.

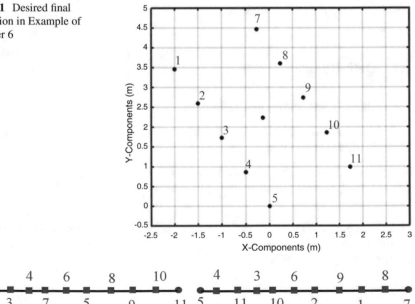

Fig. 6.1 Desired final formation in Example of Chapter 6

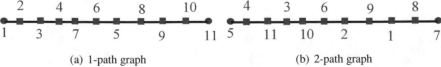

(a) 1-path graph

(b) 2-path graph

Fig. 6.2 q-paths (q=1, 2) used in example of Chapter 6

Table 6.1 Agents' positions and q-index numbers that are consistent with the configuration shown in Fig. 6.1

Order Number	1st coordinate ($q = 1$)			2nd coordinate ($q = 2$)		
	$x_{f,i,1}$	1-index	Type	$x_{f,i,2}$	2-index	Type
1	−2.00	1	1-Leader	0.00	5	2-Leader
2	1.7321	11	1-Leader	4.4641	7	2-Leader
3	−1.5000	2	1-Follower	0.8660	4	2-Follower
4	−1.0000	3	1-Follower	1.0000	11	2-Follower
5	−0.5000	4	1-Follower	1.7321	3	2-Follower
6	−0.2679	7	1-Follower	1.8660	10	2-Follower
7	−0.1340	6	1-Follower	2.2321	6	2-Follower
8	0	5	1-Follower	2.5981	2	2-Follower
9	0.2321	8	1-Follower	2.7321	9	2-Follower
10	0.7321	9	1-Follower	3.4641	1	2-Follower
11	1.2321	10	1-Follower	3.5981	8	2-Follower

Definition 6.6. Let $ql_1, ql_2 \in V$ be index numbers of the two q-leaders guiding the motion of the qth components of the q-followers' positions, then

$$V_{lq} = \{ql_1, ql_2\}$$

is called the *q-leader set*.

Definition 6.7. Index numbers of the q-followers are defined by the set

$$V_{fq} = V \backslash V_l q.$$

The set V_{fq} is called the *q-follower set*.

Definition 6.8. The set $N_{q,i}$, called *q-in-neighbor set of the q-agent i*, defines the index numbers of the two adjacent q-agents $i_{1,q}$ and $i_{1,q}$ whose states can be accessed by the q-follower i.

Definition 6.9. The vector

$$O = [1 \dots N]^T \in \mathbb{R}^N, \tag{6.1}$$

called the *order number vector*, defines order numbers of the q-agents.

Remark 6.1. It is noted that order numbers of the two q-leaders are $r = 1$ and $r = 2$. Order numbers of the q-followers are increasing from left to right, where the order numbers of the leftmost and the rightmost q-followers are 3 and N, respectively.

Definition 6.10. The order number vector $O \in \mathbb{R}^N$ can be uniquely related to the q-index number vector $In^q \in \mathbb{R}^N$ by

$$In^q = P_q O, \quad q = 1, 2, , n, \tag{6.2}$$

where $P_q \in \mathbb{R}^{N \times N}$ is an orthogonal matrix. The rs entry of P_q is given by

$$P_{q_{rs}} = \begin{cases} 1 & s = i(r) \\ 0 & Otherwise \end{cases} \tag{6.3}$$

As observed, there is one-to-one mapping between the q-index numbers and the order number vector. In other words, r (the rth entry of the vector O) denotes the order number of the q-agent whose q-index is r.

The vectors In^1 and In^2 as well as the matrices P_1 and P_2, corresponding to the desired formation shown in Fig. 6.1 and the communication q-paths shown in Fig. 6.2, are obtained as follows:

$$In^1 = \begin{bmatrix} 1 & 11 & 2 & 3 & 4 & 7 & 6 & 5 & 8 & 9 & 10 \end{bmatrix}^T \in \mathbb{R}^{11} \tag{6.4}$$

$$In^2 = \begin{bmatrix} 5 & 7 & 4 & 11 & 3 & 10 & 6 & 2 & 9 & 1 & 8 \end{bmatrix}^T \in \mathbb{R}^{11} \tag{6.5}$$

$$P_1 = \begin{bmatrix} 1 & 0 & 0 & 0 & 0 & 0 & 0 & 0 & 0 & 0 & 0 \\ 0 & 0 & 0 & 0 & 0 & 0 & 0 & 0 & 0 & 0 & 1 \\ 0 & 1 & 0 & 0 & 0 & 0 & 0 & 0 & 0 & 0 & 0 \\ 0 & 0 & 1 & 0 & 0 & 0 & 0 & 0 & 0 & 0 & 0 \\ 0 & 0 & 0 & 1 & 0 & 0 & 0 & 0 & 0 & 0 & 0 \\ 0 & 0 & 0 & 0 & 0 & 1 & 0 & 0 & 0 & 0 & 0 \\ 0 & 0 & 0 & 0 & 0 & 1 & 0 & 0 & 0 & 0 & 0 \\ 0 & 0 & 0 & 0 & 1 & 0 & 0 & 0 & 0 & 0 & 0 \\ 0 & 0 & 0 & 0 & 0 & 0 & 0 & 1 & 0 & 0 & 0 \\ 0 & 0 & 0 & 0 & 0 & 0 & 0 & 0 & 1 & 0 & 0 \\ 0 & 0 & 0 & 0 & 0 & 0 & 0 & 0 & 0 & 1 & 0 \end{bmatrix}^T \in \mathbb{R}^{11 \times 11} \tag{6.6}$$

$$P_2 = \begin{bmatrix} 0 & 0 & 0 & 0 & 1 & 0 & 0 & 0 & 0 & 0 & 0 \\ 0 & 0 & 0 & 0 & 0 & 0 & 1 & 0 & 0 & 0 & 0 \\ 0 & 0 & 0 & 1 & 0 & 0 & 0 & 0 & 0 & 0 & 0 \\ 0 & 0 & 0 & 0 & 0 & 0 & 0 & 0 & 0 & 0 & 1 \\ 0 & 0 & 1 & 0 & 0 & 0 & 0 & 0 & 0 & 0 & 0 \\ 0 & 0 & 0 & 0 & 0 & 0 & 0 & 0 & 0 & 1 & 0 \\ 0 & 0 & 0 & 0 & 0 & 1 & 0 & 0 & 0 & 0 & 0 \\ 0 & 1 & 0 & 0 & 0 & 0 & 0 & 0 & 0 & 0 & 0 \\ 0 & 0 & 0 & 0 & 0 & 0 & 0 & 0 & 1 & 0 & 0 \\ 1 & 0 & 0 & 0 & 0 & 0 & 0 & 0 & 0 & 0 & 0 \\ 0 & 0 & 0 & 0 & 0 & 0 & 0 & 1 & 0 & 0 & 0 \end{bmatrix}^T \in \mathbb{R}^{11 \times 11} \tag{6.7}$$

Fig. 6.3 Schematic of an appropriate orientation of the Cartesian coordinate system

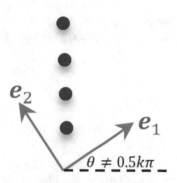

Remark 6.2. The basis of the Cartesian coordinates $(\hat{\mathbf{e}}_1, \ldots, \hat{\mathbf{e}}_n)$ are oriented such that the qth components of the positions of no two different q-agents are the same in the final desired configuration. For instance, consider the desired configuration that is shown in Fig. 6.3. If basis $\hat{\mathbf{e}}_1$ and $\hat{\mathbf{e}}_2$ are parallel to the horizontal and vertical lines, respectively, then the first components of positions of all 1-agents are the same. However, the first and second components of desired positions of no two different agents are the same, if the unit vector $\hat{\mathbf{e}}_1$ makes an angle $\theta \neq \dfrac{k\pi}{2}$ ($k \in \mathbb{Z}$) with the horizontal lines.

Remark 6.3. If the total number of agents is finite and no two agents occupy the same position, then there exists an appropriate orientation for the Cartesian coordinate system such that the qth components of no two agents are the same. For example consider a distribution of agents in the $X - Y$ plane, where N agents are arbitrarily distributed and no two agents have the same position. Suppose that θ is the angle between X axis and a horizontal line. For every pair of agents, X and Y are different if neither $\hat{\mathbf{e}}_x$ or $\hat{\mathbf{e}}_y$ is parallel to the line connecting the two agents. Therefore, four different orientations of the $X - Y$ coordinate system should be avoided for every pair of agents. This implies that θ should not take $\dfrac{4N(N-1)}{2}$ discrete values denoted by $\theta_1, \theta_2, \ldots, \theta_{2N(N-1)}$. So, none of the agents have the same q-component, if

$$\theta \in [0, 360] \setminus \{\theta_1, \ldots, \theta_{2N(N-1)}\}.$$

6.2 q-Communication Weight and q-Weight Matrix

The interagent communications among q-followers are defined by a q-path. The boundary nodes of the q-path represent two q-leaders. The interior nodes of the q-path represent q-followers. Let $x_{f,q,i}$ (the qth component of the final desired position of the q-follower i) be expressed as the linear communication of x_{f,q,i_1} and

x_{f,q,i_2} (x_{f,q,i_1} and x_{f,q,i_2} are the qth components of the final desired positions of the q-agents $i_{1,q}$ and $i_{2,q}$) as follows:

$$x_{f,q,i} = w_{i,i_1}{}^q x_{f,q,i_1} + w_{i,i_2}{}^q x_{f,q,i_2}, \tag{6.8}$$

where

$$w_{i,i_1}{}^q + w_{i,i_2}{}^q = 1. \tag{6.9}$$

Then, the parameters $w_{i,i_1}{}^q$ and $w_{i,i_2}{}^q$, called *q-communication weights*, are uniquely determined based on the qth components of the agents' positions in the final configuration by

$$\begin{cases} w_{i,i_1}{}^q = \dfrac{x_{f,q,i_2} - x_{f,q,i}}{x_{f,q,i_2} - x_{f,q,i_1}} \\ w_{i,i_2}{}^q = \dfrac{x_{f,q,i} - x_{f,q,i_1}}{x_{f,q,i_2} - x_{f,q,i_1}} \end{cases}. \tag{6.10}$$

The adjacent agents $i_{1,q}$ and $i_{2,q}$ and the q-communication weights corresponding to the desired MAS configuration shown in Fig. 6.1 are listed in Table 6.2.

q-Weight Matrix: The q-weight matrix $W_q \in \mathbb{R}^{N \times N}$ with the ij entry of W_q can be defined as follows:

$$W_{q_{i,j}} = \begin{cases} 0 & i \in V_{lq} \\ w_{i,j}{}^q & i \in V_{fq} \wedge j \in N_{q,i} \\ -1 & i \in V_{fq} \wedge i = j \\ 0 & \text{Otherwise} \end{cases}. \tag{6.11}$$

Table 6.2 Adjacent agents $i_{1,q}$ and $i_{2,q}$ and the q-communication weights that are consistent with the MAS configuration shown in Fig. 6.1

Index i	q = 1				q = 2			
	$i_{1,q}$	$i_{2,q}$	$w_{i,i_1}{}^q$	$w_{i,i_2}{}^q$	$i_{1,q}$	$i_{2,q}$	$w_{i,i_1}{}^q$	$w_{i,i_2}{}^q$
1	-	-	-	-	8	9	0.8453	0.1547
2	1	3	0.5000	0.5000	6	9	0.2679	0.7321
3	2	4	0.5000	0.5000	10	11	0.8453	0.1547
4	3	7	0.3170	0.6830	5	11	0.1340	0.8660
5	6	8	0.6340	0.3660	-	-	-	-
6	5	7	0.5000	0.5000	2	10	0.5000	0.5000
7	4	6	0.3660	0.6340	-	-	-	-
8	5	9	0.6830	0.3170	1	7	0.8660	0.1340
9	8	10	0.5000	0.5000	1	2	0.1547	0.8453
10	9	11	0.5000	0.5000	3	6	0.7321	0.2679
11	-	-	-	-	3	4	0.1547	0.8453

The following properties can be counted for the matrix W_q:

- The matrix W_q is zero-sum row (sum of the entries of each row of the matrix W_q is zero).
- All the entries of the rows ql_1 and ql_2 of the matrix W_q are zero. This is because the qth components of the positions of the q-leaders ql_1 and ql_2 evolve independently.

The matrices $W_1 \in \mathbb{R}^{11 \times 11}$ and $W_2 \in \mathbb{R}^{11 \times 11}$ corresponding to the q-communication weights listed in Table 6.2 are obtained as follows:

$$
W_1 = \begin{bmatrix}
0 & 0 & 0 & 0 & 0 & 0 & 0 & 0 & 0 & 0 & 0 \\
0.5 & -1 & 0.5 & 0 & 0 & 0 & 0 & 0 & 0 & 0 & 0 \\
0 & 0.5 & -1 & 0.5 & 0 & 0 & 0 & 0 & 0 & 0 & 0 \\
0 & 0 & 0.3170 & -1 & 0 & 0 & 0.6830 & 0 & 0 & 0 & 0 \\
0 & 0 & 0 & 0 & -1 & 0.6340 & 0 & 0.3660 & 0 & 0 & 0 \\
0 & 0 & 0 & 0 & 0.5 & -1 & 0.5 & 0 & 0 & 0 & 0 \\
0 & 0 & 0 & 0.3660 & 0 & 0.6340 & -1 & 0 & 0 & 0 & 0 \\
0 & 0 & 0 & 0 & 0.6830 & 0 & 0 & -1 & 0.3170 & 0 & 0 \\
0 & 0 & 0 & 0 & 0 & 0 & 0 & 0.5 & -1 & 0.5 & 0 \\
0 & 0 & 0 & 0 & 0 & 0 & 0 & 0 & 0.5 & -1 & 0.5 \\
0 & 0 & 0 & 0 & 0 & 0 & 0 & 0 & 0 & 0 & 0
\end{bmatrix}
$$

(6.12)

$$
W_2 = \begin{bmatrix}
-1 & 0 & 0 & 0 & 0 & 0 & 0 & 0.8453 & 0.1547 & 0 & 0 \\
0 & -1 & 0 & 0 & 0 & 0.2679 & 0 & 0 & 0.7321 & 0 & 0 \\
0 & 0 & -1 & 0 & 0 & 0 & 0 & 0 & 0 & 0.8453 & 0.1547 \\
0 & 0 & 0 & -1 & 0.1340 & 0 & 0 & 0 & 0 & 0 & 0.8660 \\
0 & 0 & 0 & 0 & 0 & 0 & 0 & 0 & 0 & 0 & 0 \\
0 & 0.5 & 0 & 0 & 0 & -1 & 0 & 0 & 0 & 0.5 & 0 \\
0 & 0 & 0 & 0 & 0 & 0 & 0 & 0 & 0 & 0 & 0 \\
0.8660 & 0 & 0 & 0 & 0 & 0 & 0.1340 & -1 & 0 & 0 & 0 \\
0.1547 & 0.8453 & 0 & 0 & 0 & 0 & 0 & 0 & -1 & 0 & 0 \\
0 & 0 & 0.7321 & 0 & 0 & 0.2679 & 0 & 0 & 0 & -1 & 0 \\
0 & 0 & 0.1547 & 0.8453 & 0 & 0 & 0 & 0 & 0 & 0 & 0
\end{bmatrix}
$$

(6.13)

Theorem 6.1. *By using*

$$
\hat{W}_q = P_q W_q P_q^T,
\tag{6.14}
$$

the matrix W_q is transformed to

$$
\hat{W}_q = \begin{bmatrix} \mathbf{0}_{2 \times 2} & \mathbf{0}_{2 \times (N-2)} \\ B_q & A_q \end{bmatrix}
\tag{6.15}
$$

where $B_q \in \mathbb{R}^{(N-2)\times 2}$ is nonnegative, and $A_q \in \mathbb{R}^{(N-2)\times(N-2)}$ is a Hurwitz matrix.

Proof. By applying the similarity transformation (6.14), $\hat{W}_q$ is obtained as given in Eq. (6.15). Let r and s denote order numbers of the q-followers i and j, respectively, then the ij entry of the matrix W_q is the same as the rs entry of the matrix $\hat{W}_q$. For example, 2-follower $i = 8$ interacts with the 2-follower $j = 1$ with the communication weight $W_{28,1} = w_{8,1}{}^2 = 0.8660$ (See Fig. 6.2 and Table 6.2.), where the order number of the 2-followers 8 is $r = 10$ and the order number of 2-follower 1 is $s = 9$ (See Fig. 6.2 and Table 6.1.) Therefore, $\hat{W}_{2_{10,9}} = W_{28,1} = 0.8660$. The entries of the first two rows of $\hat{W}_q$ are all zero because the order numbers of q-leaders are 1 and 2, and the qth components of the q-leaders' positions evolve independently. It is noted that the sum of each row of the matrix $\hat{W}_q$ is zero because pre- and post-multiplying of W_q by P_q and $P_q{}^T$ do not change the sum of each row of W_q that is zero. The matrix $B_q \in \mathbb{R}^{(N-2)\times 2}$ has only two positive entries $B_{q_{1,1}}$ and $B_{q_{(N-2),2}}$, where the remaining entries of the partition B_q are all zero. This is because the q-followers with the order number 3 and N communicate with the q-leaders with the order number 1 and 2, respectively. It is also noticed that all diagonal elements of A_q are -1, where entries of the matrix W_q are defined by Eq. (6.12). Therefore,

$$A_q = -(I - F_q), \tag{6.16}$$

where $F_q \in \mathbb{R}^{(N-2)\times(N-2)}$ is irreducible and nonnegative matrix; the nonzero entries of F is defined by

$$\begin{cases} F_{q_{r,r+1}} = w_{i(r+2),i(r+3)}{}^q & \text{if } r = 1 \\ F_{q_{r,r-1}} = w_{i(r+2),i(r+1)}{}^q, \; F_{q_{r,r+1}} = w_{i(r+2),i(r+3)}{}^q & \text{if } 1 < r < N-2 \\ F_{q_{r,r-1}} = w_{i(r+2),i(r+1)}{}^q & r = N-2 \\ F_{q_{r,s}} = 0 & \text{Otherwise} \end{cases} \tag{6.17}$$

Notice that $F_{q_{r,s}}$ is the rs entry of the matrix F_q, $w^q_{i(r+2),i(s+2)}$ is the q-communication weight of the q-agent $i(r+2)$ with the q-agent $i(s+2)$ ($r+2$ and $s+2$ are order numbers of the q-agents $i(r+2)$ and $i(s+2)$.). Because sum of each row of $\hat{W}_q$ is zero, sum of entries of the first row of the matrix A_q and sum of entries of the last row of the matrix A_q are both negative. The remaining rows of the matrix A_q are zero-sum. This implies that sum of the entries in the first and last rows of the matrix F_q are both less than 1, while the sum of the entries of the remaining rows of F_q is equal to 1. Therefore, by provoking Gershgorin theorem and Perron-Frobenius theorem, it is concluded that the spectral radius of F_q, $\rho(F_q)$, is less than or equal to 1. Nevertheless, the matrix $-A_q = (I - F_q)$ is a non-singular irreducible M-matrix. Consequently, $\rho(F_q)$ has to be less than 1 and A_q is inevitably Hurwitz.

The block tridiagonal matrices $\hat{W}_1$ and $\hat{W}_2$ corresponding to the desired formation shown in Fig. 6.1 and communication weights listed in Table 6.2 are obtained as follows:

$$\hat{W}_1 = \begin{bmatrix} 0 & 0 & 0 & 0 & 0 & 0 & 0 & 0 & 0 & 0 & 0 \\ 0 & 0 & 0 & 0 & 0 & 0 & 0 & 0 & 0 & 0 & 0 \\ 0.5000 & 0 & -1 & 0.5000 & 0 & 0 & 0 & 0 & 0 & 0 & 0 \\ 0 & 0 & 0.5000 & -1 & 0.5000 & 0 & 0 & 0 & 0 & 0 & 0 \\ 0 & 0 & 0 & 0.3170 & -1 & 0.6830 & 0 & 0 & 0 & 0 & 0 \\ 0 & 0 & 0 & 0 & 0.3660 & -1 & 0.6340 & 0 & 0 & 0 & 0 \\ 0 & 0 & 0 & 0 & 0 & 0.5000 & -1 & 0.5000 & 0 & 0 & 0 \\ 0 & 0 & 0 & 0 & 0 & 0 & 0.6340 & -1 & 0.3660 & 0 & 0 \\ 0 & 0 & 0 & 0 & 0 & 0 & 0 & 0.6830 & -1 & 0.3170 & 0 \\ 0 & 0 & 0 & 0 & 0 & 0 & 0 & 0 & 0.5000 & -1 & 0.5000 \\ 0 & 0.5000 & 0 & 0 & 0 & 0 & 0 & 0 & 0 & 0.5000 & -1 \end{bmatrix}$$

$$(6.18)$$

$$\hat{W}_2 = \begin{bmatrix} 0 & 0 & 0 & 0 & 0 & 0 & 0 & 0 & 0 & 0 & 0 \\ 0 & 0 & 0 & 0 & 0 & 0 & 0 & 0 & 0 & 0 & 0 \\ 0.1340 & 0 & -1 & 0.5000 & 0.8660 & 0 & 0 & 0 & 0 & 0 & 0 \\ 0 & 0 & 0.8453 & -1 & 0.1547 & 0 & 0 & 0 & 0 & 0 & 0 \\ 0 & 0 & 0 & 0.1547 & -1 & 0.8453 & 0 & 0 & 0 & 0 & 0 \\ 0 & 0 & 0 & 0 & 0.7321 & -1 & 0.2679 & 0 & 0 & 0 & 0 \\ 0 & 0 & 0 & 0 & 0 & 0.5000 & -1 & 0.5000 & 0 & 0 & 0 \\ 0 & 0 & 0 & 0 & 0 & 0 & 0.2679 & -1 & 0.7321 & 0 & 0 \\ 0 & 0 & 0 & 0 & 0 & 0 & 0 & 0.8453 & -1 & 0.1547 & 0 \\ 0 & 0 & 0 & 0 & 0 & 0 & 0 & 0 & 0.1547 & -1 & 0.8453 \\ 0 & 0.1340 & 0 & 0 & 0 & 0 & 0 & 0 & 0 & 0.8660 & -1 \end{bmatrix}$$

$$(6.19)$$

6.3 Agents' positions in the Desired Configuration

Let

$$Z_{F,q} = \begin{bmatrix} x_{f,q,1}) & \dots & x_{f,q,N} \end{bmatrix}^T \in \mathbb{R}^N$$

be the qth components of the agents' positions in the final desired configuration, then,

$$W_q Z_{F,q} = 0. \tag{6.20}$$

It is noted that the row i of Eq. (6.20) is the same as Eq. (6.8), where q-communication weights $w_{i,i_1}{}^q$ and $w_{i,i_2}{}^q$ are uniquely determined by Eq. (6.10). Let

$$\hat{Z}_{F,q} = P_q Z_{F,q}, \tag{6.21}$$

then the rth entry of $\hat{Z}_{F,q} \in \mathbb{R}^N$ is the equal to $x_{f,q,i}$ (the ith entry of $Z_{F,q}$), where r is the order number of the q-agent i. Now, $Z_{F,q}$ and W_q in Eq. (6.20) can be replaced by $Z_{F,q} = P_q^{-1}\hat{Z}_{F,q}$ and $W_q = P_q^T \hat{W}_q P_q$, respectively, then

$$\hat{W}_q \hat{Z}_q = A_q f_{F,q} + B_k l_{L,q} = 0. \tag{6.22}$$

It is noted that $\hat{Z}_{F,q}$ in Eq. (6.22) is partitioned as

$$\hat{Z}_{F,q} = \begin{bmatrix} l_{L,q} \\ f_{F,q} \end{bmatrix} \tag{6.23}$$

where $l_{F,q} \in \mathbb{R}^2$ (the first two rows of $\hat{Z}_{F,q}$ represents the qth components of the final positions of the q-leaders, while $f_{F,q} \in \mathbb{R}^{N-2}$ (the last N-2 rows of $\hat{Z}_{F,q}$) specifies the qth components of the followers' desired final positions.

Transient Desired Positions of the q-followers: Let $x_{f,q,i}$ (the qth component of the final desired position of the q-follower i) be expanded as the linear combination of x_{f,q,ql_1} and x_{f,q,ql_2} (the qth components of the ultimate positions of the two q-leaders identified by the index numbers ql_1 and ql_2) as follows:

$$x_{f,q,i} = \alpha_{i,1}{}^q x_{f,q,ql_1} + \alpha_{i,2}{}^q x_{f,q,ql_2}, \tag{6.24}$$

where

$$\alpha_{i,1}{}^q + \alpha_{i,2}{}^q = 1. \tag{6.25}$$

Then, $\alpha_{i,1}{}^q$ and $\alpha_{i,2}{}^q$ are uniquely determined as follows:

$$\begin{cases} \alpha_{i,1}{}^q = \dfrac{x_{f,q,ql_2} - x_{f,q,i}}{x_{f,q,ql_2} - x_{f,q,ql_1}} \\ \alpha_{i,2}{}^q = \dfrac{x_{f,q,i} - x_{f,q,ql_1}}{x_{f,q,ql_2} - x_{f,q,ql_1}} \end{cases}. \tag{6.26}$$

The parameters $\alpha_{i,1}{}^q$ and $\alpha_{i,2}{}^q$ that are consistent with the agents' desired final positions are listed in Table 6.3. The qth component of the transient desired position of the q-follower i, given by

$$x_{d,q,i}(t) = \alpha_{i,1}{}^q x_{(q,ql_1)}(t) + \alpha_{i,2}{}^q x_{(q,ql_2)}(t). \tag{6.27}$$

Table 6.3 Parameters $\alpha_{i,1}{}^q$ and $\alpha_{i,2}{}^q$ corresponding to the desired formation shown in Fig. 6.1

	$x_{f,1,i}$	Index $i(r)$	$\alpha_{i,1}{}^1$	$\alpha_{i,2}{}^1$	$x_{f,2,i}$	Index $i(r)$	$\alpha_{i,1}{}^2$	$\alpha_{i,2}{}^2$
1	−2.00	1	1	0	0.00	5	1	0
2	1.7321	11	0	1	4.4641	7	0	1
3	−1.5000	2	0.8660	0.1340	0.8660	4	0.8060	0.1940
4	−1.0000	3	0.7321	0.2679	1.0000	11	0.7760	0.2240
5	−0.5000	4	0.5981	0.4019	1.7321	3	0.6120	0.3880
6	−0.2679	7	0.5359	0.4641	1.8660	10	0.5820	0.4180
7	−0.1340	6	0.5000	0.5000	2.2321	6	0.5000	0.5000
8	0	5	0.4641	0.5359	2.5981	2	0.4180	0.5820
9	0.2321	8	0.4019	0.5981	2.7321	9	0.3880	0.6120
10	0.7321	9	0.2679	0.7321	3.4641	1	0.2240	0.7760
11	1.2321	10	0.1340	0.8660	3.5981	8	0.1940	0.8060

is the same as the row $r - 2$ of

$$f_{q,d} = -A_q{}^{-1} B_q \begin{bmatrix} x_{q,ql_1} \\ x_{q,ql_2} \end{bmatrix}. \tag{6.28}$$

6.4 Dynamics of Agents

6.4.1 First-Order Dynamics for Evolution of the MAS

Assume that the q-agent $i \in V$ updates the qth component of its current position by

$$\frac{dx_{q,i}}{dt} = u_{q,i}, \tag{6.29}$$

where

$$u_{q,i} = \begin{cases} given & i \in V_{lq} \\ g\left(\sum_{j \in N_{q,i}} wi, j^q x_{q,j} - x_{q,i}\right) & i \in V_{fq} \end{cases}. \tag{6.30}$$

It is noted that the control gain $g \in \mathbb{R}_+$ is constant. Then, the qth components of the q-followers' positions are updated according to

$$\dot{z}_q = g W_q z_q + v_q, \tag{6.31}$$

where $z_q = [x_{q,1} \ldots x_{q,N}]^T \in \mathbb{R}^N$. Also, except the rows ql_1 and ql_2 of the vector $v_q \in \mathbb{R}^N$ that specifies the qth components of the q-leaders' velocities, the remaining entries of v_q are zero. If z_q, v_q, and W_q in Eq. (6.31) are replaced by $P_q{}^{-1}\hat{z}_q$, $P_q{}^{-1}\hat{v}_q$ and $P_q{}^T \hat{W}_q P_q$, respectively, then

$$\dot{\hat{z}}_q = g\hat{W}_q\hat{z}_q + \hat{v}_q = g \begin{bmatrix} \mathbf{0}_{2\times2} & \mathbf{0}_{2\times(N-2)} \\ B_q & A_q \end{bmatrix} \hat{z}_q + \begin{bmatrix} \hat{v}_{L,q} \\ \mathbf{0}_{(N-2)\times1} \end{bmatrix}. \tag{6.32}$$

It is noted that $\hat{v}_{L,q} = [v_{ql_1}) \quad v_{ql_2}]^T$ specifies the qth components of the q-leaders' velocities. Also, the MAS evolution dynamics represented by Eq. (6.32) is stable because communication weights, that are determined by Eq. (6.10), are all positive. Therefore the matrix A_q is Hurwitz. The equilibrium state of Eq. (6.32) assigns the qth components of the agents' positions in the final configuration when q-leaders ultimately stop ($\hat{v}_{L,q}$ and $\dot{\hat{z}}_q$ are both zero at the equilibrium state). Therefore, the q-followers asymptotically reach the desired final positions that satisfy Eq. (6.28).

Remark 6.4. Applying the proposed first-order kinematic model can guarantee asymptotic convergence of the agents' transient positions to the desired positions in the final configuration; however, it requires leaders to stop in a finite horizon of time. In other words, positions of the followers in the transient configuration deviate from the desired positions, although they eventually reach the desired configuration. Next, it is shown how transient deviations of followers from the desired positions (that are defined globally based on positions of the q-leaders) vanish during MAS evolution, where (i) q-followers access positions and velocities of only two adjacent q-agents, and (ii) leaders move with constant velocities.

6.4.2   Double Integrator Kinematic Model (Asymptotic Tracking of a Moving Desired Formation)

Let the q-follower $i \in V$ update the qth component of its current position by

$$\frac{d^2x_{q,i}}{dt^2} = u_{q,i}, \tag{6.33}$$

where

$$u_{q,i} = \begin{cases} given & i \in V_{lq} \\ g_1 \left(\sum_{j\in N_{q,i}} w^q_{i,j}\dot{x}_{q,j} - \dot{x}_{q,i} \right) + g_2 \left(\sum_{j\in N_{q,i}} w^q_{i,j}x_{q,j} - x_{q,i} \right) & i \in V_{fq} \end{cases}. \tag{6.34}$$

The qth components of the q-followers' positions are then updated by the following second order matrix dynamics:

$$\ddot{z}_q = g_1 W_q\dot{z}_q + g_2 W_q z_q + a_q, \tag{6.35}$$

Note that z_q and W_q were previously introduced, and $g_1, g_2 \in \mathbb{R}_+$ and are constant. Except the rows ql_1 and ql_2 of the vector $a_q \in \mathbb{R}^N$, that represent the qth components of the q-leaders' accelerations, the remaining entries of a_q are zero. Replacing $z_q = P_q^{-1}\hat{z}_q$, $\dot{z}_q = P_q^{-1}\dot{\hat{z}}_q$, $a_q = P_q^{-1}\hat{a}_q$, and $W_q = P_q^T \hat{W}_q P_q$, results in

$$\ddot{\hat{z}}_q = g_1 \hat{W}_q \dot{\hat{z}}_q + g_2 \hat{W}_q \hat{z}_q + \hat{a}_q, \tag{6.36}$$

where $\hat{a}_q = [\hat{a}_{L,q} \quad \mathbf{0}_{1\times(N-2)}]^T \in \mathbb{R}^N$, and $\hat{a}_{L,q} = [a_{ql_1} \quad a_{ql_2}]^T$ defines the qth components of the q-leaders' accelerations. Let

$$\hat{z}_q = \begin{bmatrix} l_q \\ f_q \end{bmatrix} \tag{6.37}$$

be substituted in Eq. (6.36), where $l_q \in \mathbb{R}^2$ denotes the qth components of the leaders' positions and $f_q \in \mathbb{R}^{N-2}$ denotes the qth components of the followers' positions. Then, f_q is updated by the following second order dynamics:

$$\ddot{f}_q = A_q(g_1\dot{f}_q + g_2 f_q) + B_q(g_1 l_q + g_2 l_q) + \hat{a}_q. \tag{6.38}$$

It is assumed that the two q-leaders move with constant velocities, therefore, $\hat{a}_q = \mathbf{0}$, and Eq. (6.38) simplifies to

$$\begin{aligned} \ddot{f}_q &= A_q(g_1\dot{f}_q + g_2 f_q) - A_q(g_1(-A_q^{-1}B_q\dot{l}_q) + g_2(-A_q^{-1}B_q l_q)) \\ &= A_q(g_1(\dot{f}_q - \dot{f}_{q,d}) + g_2(f_q - f_{q,d})). \end{aligned} \tag{6.39}$$

Because $\hat{a}_q = \mathbf{0}$ ($\ddot{l}_q = [\ddot{x}_{q,lq_1} \quad \ddot{x}_{q,lq_2}]^T = \mathbf{0}$), $\ddot{f}_{q,d} = -A^{-1}B\ddot{l}_q = 0$. Therefore, the transient error $E_q = f_q - f_{q,d}$ is updated by the following second order linear dynamics:

$$\ddot{E}_q - A_q(g_1\dot{E}_q + g_2 E_q) = 0, \tag{6.40}$$

where the roots of the characteristic polynomial

$$|s^2 I - (g_1 s + g_2)A_q| = 0 \tag{6.41}$$

are all located in the open left half s-plane. Consequently the transient error converges to zero during MAS evolution, while the q-followers do not directly access the qth components of positions and velocities of q-leaders.

Example: Consider the MAS containing 11 agents, where the index numbers and the final desired positions of the agents are listed in Table 6.1. Agents' initial positions are shown in Fig. 6.4.

It is aimed that the agents reach the desired formation shown in Fig. 6.1. For this purpose, the first components of agents' positions are independently guided by the

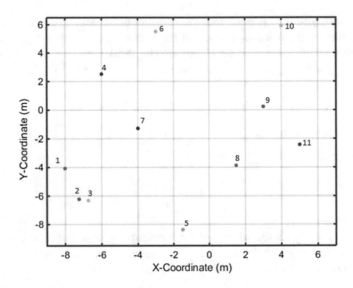

Fig. 6.4 Initial distribution of the agents in Example of the Chapter 6

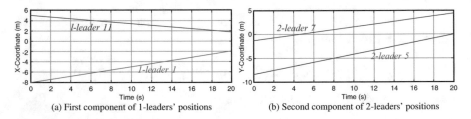

(a) First component of 1-leaders' positions   (b) Second component of 2-leaders' positions

Fig. 6.5 The qth components of positions of q-leaders

1-leaders 1 and 11, where 1-followers acquire the desired position through local communication. It is noticed that followers 1-communication weights that are listed in Table 6.2. Evolution of the second components of the agents' positions are guided by the 2-leaders 5 and 7, where the 2-followers apply the 2-communication weights listed in Table 6.2 to acquire the second components of the desired positions through local interaction.

In Fig. 6.5(a), the first components of positions of the 1-leaders 1 and 11 are illustrated versus time. Also, the second components of positions of the 2-leaders 5 and 7 are shown in Fig. 6.5(b) as functions of time. As seen q-leaders move with constant velocities during the time $t \in [0, 20]s$. Note that leaders stop at t=20s.

Simulation of the evolution of the agents under the proposed first and second order models are presented below.

First Order Model: Let q-followers update the qth components of their positions according to Eq. (6.29) and Eq. (6.30), where $g_1 = g_2 = 30$. Formations of the MAS at the times $t = 0.5s, t = 1s, t = 2s, t = 7s, t = 14s$, and $t = 25s$ are shown in Fig. 6.6. As it is observed, the MAS ultimately form the letter "H" as it is desired.

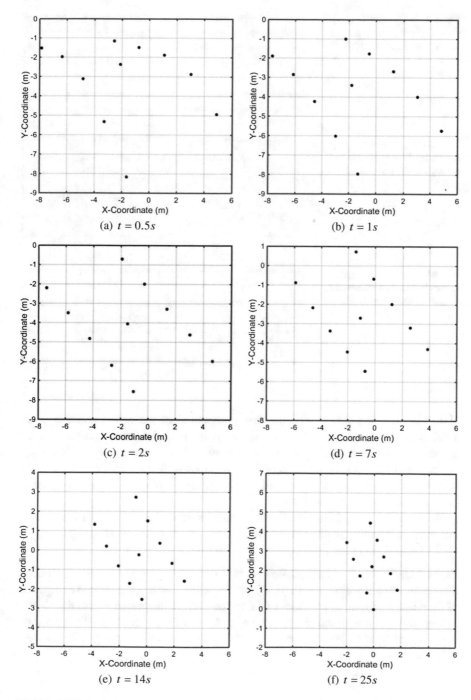

Fig. 6.6 MAS formations at six different sample times $t = 0.5s$, $t = 1s$, $t = 2s$, $t = 7s$, $t = 14s$, and $t = 25s$

Fig. 6.7 $x_{1,6}(t)$ and $x_{d,1,6}(t)$ shown by the continuous and dotted curves, respectively

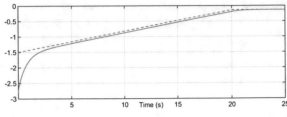

Fig. 6.8 $x_{2,6}(t)$ and $x_{d,2,6}(t)$ shown by the continuous and dotted curves, respectively

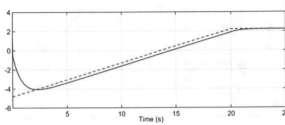

Fig. 6.9 $x_{2,6}(t)$ and $x_{d,2,6}(t)$ shown by the continuous and dotted curves, respectively

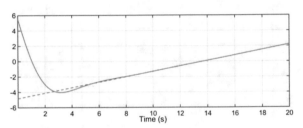

In Figs. 6.7 and 6.8, $x_{1,6}(t)$ (the first component of the actual position of the 1-follower 6) and $x_{2,6}(t)$ (the second component of the actual position of the 2-follower 6) are shown by continuous curves. Furthermore, $x_{d,1,6}(t)$ (the first component of the desired position of the 1-follower 6 (See Eq. (6.27))) and $x_{d,2,6}(t)$ (the second component of the desired position of the 2-follower 6) are illustrated by dotted curves in Figs. 6.7 and 6.8.

As observed, the q-follower 6 deviates from its desired position during transition; however, it asymptotically reaches the desired final position at $(x_{f,1,6}, x_{f,2,6}) = (-0.1340, 2.2321)$ after leaders stop at $t = 20s$.

Second Order Model: Let q-followers choose $g_1 = g_2 = 30$, where they update their positions according to Eqs. (6.33) and (6.34). In Fig. 6.9, the second component of the desired and actual positions of the follower 6 are illustrated versus time by the dashed and continuous curves, respectively. As observed, deviation of the follower 6 converges to zero around $t = 7s$, while leaders are still moving with constant velocities.

Chapter 7
Continuum Deformation of a Multi-Agent System over Nonlinear Surfaces

In this chapter, it is shown how a multi-agent system (MAS) consisting of N agents can move collectively on a nonlinear surface through local communication. Collective motion of the MAS is treated as continuum deformation, therefore, interagent distances can be largely expanded or contracted. For collective motion on a p-D ($p \leq 3$) nonlinear surface, leader-follower approach is applied and continuum deformation is prescribed by $p + 1$ leaders that move independently. Each follower uses a first-order discrete-time dynamics and communicate with $p + 1$ in-neighbor agents to acquire desired position specified by the continuum deformation. Similar to collective motion on linear surfaces, followers' communication weights are consistent with agents' initial positions. Examples of collective motion on an arbitrary curve as well as a 2-D nonlinear surface in a 3-D motion space are also provided.

7.1 Discrete-Time Dynamics of Collective Motion on a Nonlinear Surface

7.1.1 Dynamics of an Individual Agent

Suppose that position of the agent i is given by

$$r_i[K] = \sum_{q=1}^{n} x_{q,i}[K]\hat{\mathbf{e}}_q \tag{7.1}$$

where $x_{q,i} = x_{q,i}(\beta_{1,i}, \ldots, \beta_{p,i})$ $(q = 1, 2, \ldots, n)$ and $(\beta_1, \ldots, \beta_p)$ is the natural coordinate of the manifold $M_p \subset \mathbb{R}^3$. It is assumed that $(\beta_{1,i}, \ldots, \beta_{p,i})$ is a regular point and $\beta_{q,i}$ $(q = 1, 2, \ldots, p)$ is updated by the first order dynamics

© Springer International Publishing AG 2016
H. Rastgoftar, *Continuum Deformation of Multi-Agent Systems*,
DOI 10.1007/978-3-319-41594-9_7

$$\beta_{q,i}[K+1] = u_{q,i} \tag{7.2}$$

where

$$u_{q,i} = \begin{cases} \text{given} & i \in V_L \\ \beta_{q,i}[K] + g\sum_{j\in N_i} w_{i,j}(\beta_{q,j}[K] - \beta_{q,i}[K]) & i \in V_F, j \in N_i \end{cases}. \tag{7.3}$$

Given $\beta_{q,i}[K]$, obtained from Eq. (7.2), $x_{l,i}[K]$ ($l = 1,2,\ldots,n$) is updated by the following difference equation:

$$x_{l,i}[K+1] = x_{l,i}[K] + \sum_{q=1}^{p}\sum_{j\in N_i} g w_{i,j}\frac{\partial x_{l,i}}{\partial \beta_{q,i}}(\beta_{q,j}[K] - \beta_{q,i}[K]). \tag{7.4}$$

7.1.2　Interagent Communication

Communication weights w_{i,i_k} ($i \in V_F, i_k \in N_i$) are consistent with agents' initial positions, where

$$\beta_{q,i}[1] = \sum_{k=1}^{p+1} w_{i,i_k}\beta_{q,i_k}[1] \tag{7.5}$$

and

$$\sum_{k=1}^{p+1} w_{i,i_k} = 1. \tag{7.6}$$

If

$$Rank\left[\beta_{q,i_2} - \beta_{q,i_1} \cdots \beta_{q,i_{p+1}} - \beta_{q,i_1}\right] = p, \tag{7.7}$$

then, $w_{i,i_1}, w_{i,i_2}, \ldots, w_{i,i_{p+1}}$ are the unique solution of the following set of $p+1$ linear algebraic equations:

$$\begin{bmatrix} \beta_{1,i_1}[1] & \cdots & \beta_{1,i_{p+1}}[1] \\ \vdots & \ddots & \vdots \\ \beta_{p,i_1}[1] & \cdots & \beta_{p,i_{p+1}}[1] \\ 1 & \cdots & 1 \end{bmatrix} \begin{bmatrix} w_{i,i_1} \\ \vdots \\ w_{i,i_{p+1}} \end{bmatrix} = \begin{bmatrix} \beta_{1,i}[1] \\ \vdots \\ \beta_{p,i}[1] \\ 1 \end{bmatrix}. \tag{7.8}$$

7.1.3   Collective Dynamics

Given communication weights obtained from Eq. (7.8), the weigh matrix W can be set up by using Eq. (3.8). Partition A and B of the weight matrix W is obtained from Eq. (3.9). Then, collective motion of the MAS is updated by

$$z_q[K+1] = z_q[K] + g(Az_q[K] + Bu_q[K]) \tag{7.9}$$

where $u_q = [\beta_{q,1} \ \cdots \ \beta_{q,p+1}]^T \in \mathbb{R}^{p+1}$ and $z_q = [\beta_{q,p+2} \ \cdots \ \beta_{q,p+1}]^T \in \mathbb{R}^{N-p-1}$.

7.2   Continuum Deformation Along a Smooth Path in a 3-*D* Motion Space

Let position of an agent $i \in V$ be realized by

$$r_i = x_{1,i}(\beta_{1,i})\hat{e}_1 + x_{2,i}(\beta_{1,i})\hat{e}_2 + x_{3,i}(\beta_{1,i})\hat{e}_3 \tag{7.10}$$

where β_1 is the natural coordinate of a desired curve $r_i = r_i(\beta_1)$ in a 3-D motion space.

Let $V_L = \{1,2\}$ and $V_F = \{3,4,\ldots,N\}$ define index numbers of leaders and followers, respectively, and index numbers of in-neighbor agents of a follower $i \in V_F$ are given by

$$N_i = \{i_1, i_2\}$$

$$i_1 = \begin{cases} 1 & i = 3 \\ i-1 & else \end{cases}. \tag{7.11}$$

$$i_2 = \begin{cases} i+1 & i < N \\ 2 & else \end{cases}$$

Thus, partition A of the weight matrix W is a tridiagonal matrix with diagonal entries that are all -1. It is assumed that

$$\forall i \in V_F, \ \beta_{1,i_1} < \beta_{1,i} < \beta_{1,i_2}, \tag{7.12}$$

therefore, communication weights obtained from Eq. (7.8), are positive.

Example 7-1. Consider an MAS consisting of $N = 20$ agents with initial distribution $\beta_{1,i}[1]$ given in the second column of Table 1. Positions of the leaders are updated by

Table 7.1 $\beta_{1,i}[1]$ ($i \in V$); in-neighbor agents of the followers; communication weights used by the followers

i	$\beta_{1,i}[1]$	i_1	i_2	w_{i,i_1}	w_{i,i_1}
1	-0.5236	–	–	–	–
2	0.5236	–	–	–	–
3	-0.4730	1	4	0.4238	0.5762
4	-0.4357	3	5	0.4997	0.5003
5	-0.3985	4	6	0.5919	0.4081
6	-0.3446	5	7	0.4681	0.5319
7	-0.2971	6	8	0.5171	0.4829
8	-0.2462	7	9	0.4448	0.5552
9	-0.2055	8	10	0.5503	0.4497
10	-0.1557	9	11	0.4510	0.5490
11	-0.1148	10	12	0.5012	0.4988
12	-0.0736	11	13	0.5398	0.4602
13	-0.0254	12	14	0.5782	0.4218
14	0.0407	13	15	0.5919	0.4081
15	0.1366	14	16	0.5094	0.4906
16	0.2362	15	17	0.4277	0.5723
17	0.3106	16	18	0.4299	0.5701
18	0.3667	17	19	0.4515	0.5485
19	0.4129	18	20	0.5681	0.4319
20	0.4737	19	2	0.4509	0.5491

$$\text{Leader 1}: \ \beta_{1,1}[K] = \frac{\dfrac{-5\pi}{6} + \dfrac{\pi}{6}}{3000} K - \frac{\pi}{6}.$$
$$\text{Leader 2}: \ \beta_{1,2}[K] = \frac{\dfrac{-5\pi}{6} - \dfrac{\pi}{6}}{3000} K + \frac{\pi}{6}$$

(7.13)

Followers use the dynamics (7.2) to update their positions through local communication, where communication weights are consistent with the initial positions of the agents, i.e., communication weights are obtained from Eq. (7.8). In Table 7.1 Index numbers of in-neighbor agents of the followers as well as communication weights used by the followers are given.

Collective Motion on a 2-D Curve: It is desired that the MAS moves collectively along the curve

$$r = 20\cos\beta_1(1 + \cos\beta_1)\hat{\mathbf{e}}_1 + 20\sin\beta_1(1 + \cos\beta_1)\hat{\mathbf{e}}_2. \tag{7.14}$$

Each follower updates its position by using Eq. (7.4), where $g = 0.3$ and

$$\frac{\partial x_{1,i}}{\partial \beta_{1,i}} = -20(\sin \beta_{1,i} + 2\sin \beta_{1,i}\cos \beta_{1,i})$$

$$\frac{\partial x_{2,i}}{\partial \beta_{1,i}} = 20(\cos \beta_{1,i} + \cos 2\beta_{1,i}\cos \beta_{1,i}). \tag{7.15}$$

Formation of the MAS at $t = 30s$ is shown in Fig. 7.1.

Collective Motion on a 3-D Curve: It is desired that the MAS moves collectively along the curve

$$r = \sin \beta_1 \hat{\mathbf{e}}_1 + \cos \beta_1 \hat{\mathbf{e}}_2 + \beta_1 \hat{\mathbf{e}}_3. \tag{7.16}$$

Thus,

$$\frac{\partial x_{1,i}}{\partial \beta_{1,i}} = \cos \beta_{1,i}$$

$$\frac{\partial x_{2,i}}{\partial \beta_{1,i}} = -\sin \beta_{1,i} \tag{7.17}$$

$$\frac{\partial x_{3,i}}{\partial \beta_{1,i}} = 1$$

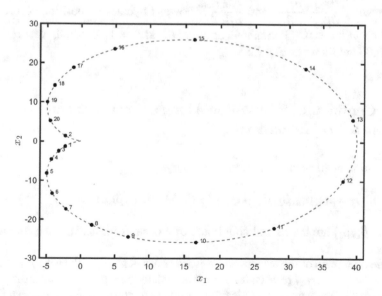

Fig. 7.1 Formation of the MAS at $t = 30s$ when $r_d = 20\cos \beta_1(1 + \cos \beta_1)\hat{e}_1 + 20\sin \beta_1(1 + \cos \beta_1)\hat{e}_2$

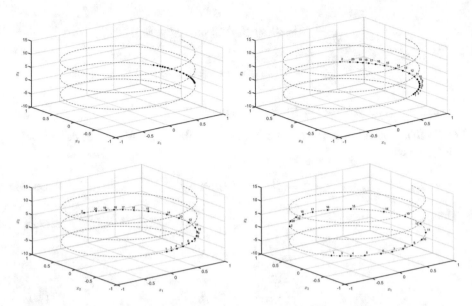

Fig. 7.2 MAS formations along the curve $r_d = \cos \beta_1 \hat{e}_1 + \sin \beta_1 \hat{e}_2 + \beta_1 \hat{e}_3$ at $K = 1$, $K = 800$, $K = 1600$, and $K = 3000$

By knowing $\dfrac{\partial x_{1,i}}{\partial \beta_{1,i}}$, $\dfrac{\partial x_{2,i}}{\partial \beta_{1,i}}$, and $\dfrac{\partial x_{3,i}}{\partial \beta_{1,i}}$, follower i updates its position according to Eq. (7.4) with $g = 0.3$. Formations of the MAS at $K = 1$, $K = 800$, $K = 1600$, and $K = 3000$ are shown in Fig. 7.2.

7.3   Continuum Deformation Along a Smooth Surface in a 3-*D* Motion Space

Let position of an agent $i \in V$ be expressed by

$$r_i = x_{1,i}(\beta_{1,i}, \beta_{2,i})\hat{e}_1 + x_{2,i}(\beta_{1,i}, \beta_{2,i})\hat{e}_2 + x_{3,i}(\beta_{1,i}, \beta_{2,i})\hat{e}_3 \qquad (7.18)$$

where (β_1, β_2) are the natural coordinate of a desired surface M_2 in a 3-*D* motion space.

Index numbers of the leaders and followers are defined by the set $V_L = \{1, 2, 3\}$ and $V_F = \{4, \ldots, N\}$, respectively. Agents update their positions according to the dynamics (7.4), where communication weights are all positive and consistent with initial distribution of agents. In example below, it is shown how an MAS can move collectively on surfaces in a 3-*D* motion space.

Fig. 7.3 Communication
graph defining interagent
communication among agents
in Example 7-2

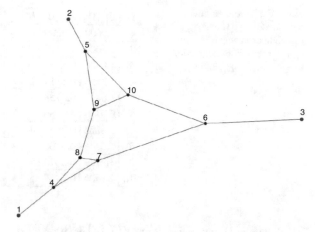

Example 7-2. Consider an MAS consisting of $N = 10$ agents (3 leaders and 7 followers) move collectively on a surface M_2 with natural coordinates β_1 and β_2. Let leaders update their positions by

$$
Leader\ 1:
\begin{cases}
\beta_{1,1}[K] = \dfrac{2.5 - 1}{3000}K + 1 \\[2ex]
\beta_{2,1}[K] = \dfrac{0.5\pi}{3000} + K
\end{cases}
$$

$$
Leader\ 2:
\begin{cases}
\beta_{1,2}[K] = \dfrac{2 - 1.5}{3000}K + 1.5 \\[2ex]
\beta_{2,2}[K] = \dfrac{\dfrac{4\pi}{3} - \dfrac{\pi}{3}}{3000}K + \dfrac{\pi}{3}
\end{cases}
\qquad (7.19)
$$

$$
Leader\ 3:
\begin{cases}
\beta_{1,3}[K] = \dfrac{6 - 4}{3000}K + 4 \\[2ex]
\beta_{2,3}[K] = \dfrac{\dfrac{4\pi}{3} - \dfrac{\pi}{6}}{3000}K + \dfrac{\pi}{6}
\end{cases}
$$

Interagent communication is defined by the graph shown in Fig. 7.3. In Table 7.2, initial natural coordinates of the agents, in-neighbor agents of the followers, and followers' communication weights are given. To acquire continuum deformation, followers update their position according to the dynamics (7.4) with $g = 0.3$.

Collective Motion on a Linear Surface : Let

$$
r_i = (2\beta_{1,i} + \beta_{2,i})\hat{e}_1 + (\beta_{1,i} + 3\beta_{2,i})\hat{e}_2 + (\beta_{1,i} + \beta_{2,i})\hat{e}_3 \qquad (7.20)
$$

define position on a desired surface in a 3-*D* motion space. When followers update their positions according to the dynamics (7.4), collective dynamics becomes as follows:

Table 7.2 Initial natural coordinates $\beta_{1,i}[1]$ and $\beta_{2,i}[1]$; in-neighbor agents i_1, i_2, and i_3; followers' communication weights w_{i,i_1}, w_{i,i_1}, and w_{i,i_3}

i	$\beta_{1,i}[1]$	$\beta_{2,i}[1]$	i_1	i_2	i_3	w_{i,i_1}	w_{i,i_1}	w_{i,i_3}
1	1.0000	0.0000	–	–	–	–	–	–
2	1.5000	1.0472	–	–	–	–	–	–
3	4.0000	0.5236	–	–	–	–	–	–
4	1.3636	0.1514	1	7	8	0.5000	0.2500	0.2500
5	1.6811	0.8758	2	9	10	0.6100	0.1900	0.2000
6	2.9720	0.4986	3	7	10	0.4900	0.2500	0.2600
7	1.8222	0.2957	4	6	8	0.3400	0.2100	0.4500
8	1.6321	0.3100	4	7	9	0.3900	0.3500	0.2600
9	1.7789	0.5672	5	8	10	0.3000	0.4400	0.2600
10	2.1404	0.6464	5	6	9	0.3300	0.3300	0.3400

$$r_i[K+1] = r_i[K] + g \sum_{j \in N_i} w_{i,j}(\beta_{1,j}[K] - \beta_{1,i}[K])(2\hat{e}_1 + \hat{e}_2 + \hat{e}_3) +$$
$$g \sum_{j \in N_i} w_{i,j}(\beta_{2,j}[K] - \beta_{2,i}[K])(\hat{e}_1 + 3\hat{e}_2 + \hat{e}_3) \qquad (7.21)$$

Shown in Fig. 7.4 are formations of the MAS when followers use the dynamics (7.21) to update their positions.

Collective Motion on a Sphere: Let

$$r_i = 10\sin\beta_{1,i}\cos\beta_{2,i}\hat{e}_1 + 10\sin\beta_{1,i}\sin\beta_{2,i}\hat{e}_2 + 10\cos\beta_{1,i} \qquad (7.22)$$

define the desired surface for collective motion of the MAS. Equation (7.22) assigns a sphere with radius 10 which is centered at the origin. Continuum deformation of the MAS on the sphere is illustrated in Fig. 7.5, by showing MAS formations at $K = 1$, $K = 1000$, $K = 2000$, and $K = 3000$.

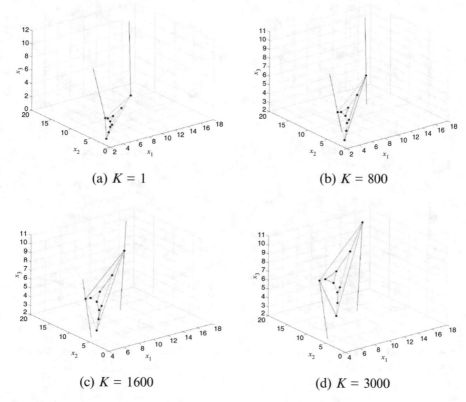

(a) $K = 1$ (b) $K = 800$

(c) $K = 1600$ (d) $K = 3000$

Fig. 7.4 Continuum deformation of an MAS on a linear surface; MAS formations at $K = 1$, $K = 800$, $K = 1600$, and $K = 3000$

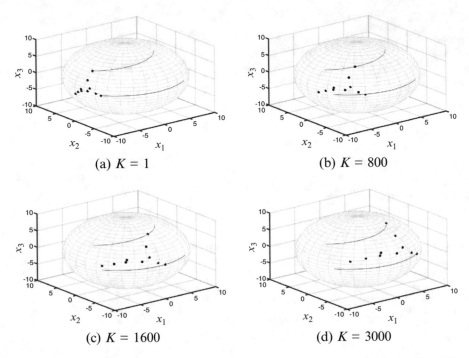

(a) $K = 1$

(b) $K = 800$

(c) $K = 1600$

(d) $K = 3000$

Fig. 7.5 Continuum deformation of an MAS on a sphere; MAS formations at $K = 1$, $K = 800$, $K = 1600$, and $K = 3000$

Appendix A
Kinematics of a Continuum

A continuous domain with an infinite number of particles is called a continuum or deformable body. A continuum deformation, denoted by the mapping $r : (\mathbb{R}^n, t) \to (\mathbb{R}^n, t)$, has the following properties:

- It is homeomorphic, i.e., a mapping $r = r(R,t)$ is homeomorphic, if the Jacobian

$$Q(R,t) = \frac{\partial r}{\partial R} \in \mathbb{R}^{n \times n} \tag{A.1}$$

is nonsingular.
- $r(R, t_0) = R$, where $R \in \mathbb{R}^n$ denotes positions of material particles in the initial formation. Initial positions of the particles are called *material coordinates*.
- $Q(R, t_0) = I_n$, where t_0 denotes the initial time and $I_n \in \mathbb{R}^{n \times n}$ is the identity matrix.
- Eigenvalues of the Jacobian matrix Q are all positive.

Under a continuum deformation, no two material particles occupy the same position. This is the key feature of a continuum deformation that is originated from the non-singularity of the Jacobian matrix.

Material and Spatial Derivatives: Let

$$R = X_1 \hat{e}_1 + X_2 \hat{e}_2 + X_3 \hat{e}_3 \tag{A.2}$$

denote initial positions of particles expressed with respect to the Cartesian coordinate system with unit basis $(\hat{e}_1, \hat{e}_2, \hat{e}_3)$. Then, position of the current configuration of the continuum is denoted by

$$r(t) = x_1(X_1, X_2, X_3, t)\hat{e}_1 + x_2(X_1, X_2, X_3, t)\hat{e}_2 + x_3(X_1, X_2, X_3, t)\hat{e}_3 \tag{A.3}$$

© Springer International Publishing AG 2016
H. Rastgoftar, *Continuum Deformation of Multi-Agent Systems*,
DOI 10.1007/978-3-319-41594-9

Suppose ϕ is a scalar function defined over position field of a continuum. If ϕ is expressed over initial position filed ($\phi = \phi(X_1, X_2, X_3, t)$), then the expression is called *material description*. On the other hand, expression of ϕ is called *spatial description*, if ϕ is expressed over the position field at the current time t ($\phi = \phi(x_1, x_2, x_3, t)$).

If the material description of ϕ is available, then,

$$\frac{D\phi}{Dt} = \frac{\partial \phi(X_1, X_2, X_3, t)}{\partial t} \tag{A.4}$$

is called *material derivative*. Furthermore, if the spatial description of ϕ is available, then,

$$\frac{D\phi}{Dt} = \frac{\partial \phi(x_1, x_2, x_3, t)}{\partial t} + \sum_{i=1}^{3} v_i(x_1, x_2, x_3, t) \frac{\partial \phi(x_1, x_2, x_3, t)}{\partial x_i} \tag{A.5}$$

is called *spatial derivative*.

Velocity and Acceleration Field of a continuum: Given position field $r = r(R, t)$, material description of the velocity field of the continuum is expressed by

$$V = \frac{Dr}{Dt} = \sum_{i=1}^{3} v_i(X_1, X_2, X_3, t) \hat{e}_i = \sum_{i=1}^{3} \frac{dx_i}{dt} \hat{e}_i. \tag{A.6}$$

Given material description of the velocity field, material description of the acceleration field is obtained as follows:

$$a = \frac{\partial V}{\partial t} = \sum_{i=1}^{3} a_i(X_1, X_2, X_3, t) \hat{e}_i = \sum_{i=1}^{3} \frac{\partial v_i(X_1, X_2, X_3, t)}{\partial t} \hat{e}_i. \tag{A.7}$$

Furthermore, spatial description of the acceleration field is expressed as follows:

$$a = \frac{DV}{Dt} = \sum_{i=1}^{3} \frac{\partial v_i(x_1, x_2, x_3, t)}{\partial t} + \sum_{i=1}^{3} v_i(x_1, x_2, x_3, t) \frac{\partial v_i(x_1, x_2, x_3, t)}{\partial x_i}. \tag{A.8}$$

Deformation Gradient Tensor: Deformation gradient tensor (matrix) of continuum deformation is defined by

$$Q = \nabla_X x = \begin{bmatrix} \dfrac{\partial x_1}{\partial X_1} & \dfrac{\partial x_1}{\partial X_2} & \dfrac{\partial x_1}{\partial X_3} \\ \dfrac{\partial x_2}{\partial X_1} & \dfrac{\partial x_2}{\partial X_2} & \dfrac{\partial x_2}{\partial X_3} \\ \dfrac{\partial x_3}{\partial X_1} & \dfrac{\partial x_3}{\partial X_2} & \dfrac{\partial x_3}{\partial X_3} \end{bmatrix}. \tag{A.9}$$

It is noted that the infinitesimal vectors $dX \in \mathbb{R}^3$ and $dx \in \mathbb{R}^3$ are related by

$$dx = QdX. \tag{A.10}$$

Polar Decomposition: The matrix Q is the Jacobian of a continuum deformation. By using polar decomposition, Q can be expressed as

$$Q = R_O U_D = V_D R_O. \tag{A.11}$$

It is noted R_O is an orthogonal matrix, U_D and V_D are symmetric positive definite matrices having the same eigenvalues. Furthermore,

$$V_D = R_O U_D R_O^T. \tag{A.12}$$

$$U_D{}^2 = Q^T Q. \tag{A.13}$$

Graphical representation of the decomposition $Q = R_O U_D$ in the $(X - Y)$ plane is shown in Fig. A.1. As it is observed, $(\hat{\mathbf{e}}_x, \hat{\mathbf{e}}_y)$ is the basis of the plane $X - Y$ while $\mathbf{e}'_x$ and $\mathbf{e}'_y)$ are orthogonal unit vectors to the disk. Also, $\mathbf{n}_1$ and $\mathbf{n}_2$ are the eigenvectors of the stretch matrix U_D. It is noted that $\mathbf{n}_1 = \mathbf{e}'_x$ and $\mathbf{n}_2 = \mathbf{e}'_y$, if the matrix U_D is diagonal.

Example A.3. Let

$$Q = \begin{bmatrix} 5 & 3 \\ 1 & 4 \end{bmatrix},$$

then,

$$U_D{}^2 = Q^T Q = \begin{bmatrix} 26 & 19 \\ 19 & 25 \end{bmatrix}.$$

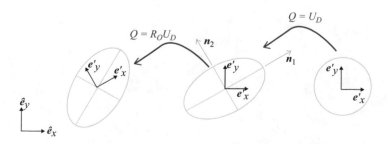

Fig. A.1 Graphical representation of polar decomposition $Q = R_O U_D$

Also, $\lambda_1(U_D{}^2) = 6.4934$ and $\lambda_2(U_D{}^2) = 44.5066$ are the eigenvalues of the matrix $U_D{}^2$. In addition, $\mathbf{n}_1 = -0.7163\hat{\mathbf{e}}_x - 0.6977\hat{\mathbf{e}}_y$ and $\mathbf{n}_2 = -0.6977\hat{\mathbf{e}}_x + 0.7163\hat{\mathbf{e}}_y$ are the eigenvectors of the matrix U_D. Eigenvectors of the matrices U_D and $U_D{}^2$ are the same, thus

$$
U_D = \begin{bmatrix} \mathbf{n}_1 & \mathbf{n}_2 \end{bmatrix} \begin{bmatrix} \sqrt{\lambda_1(U_D{}^2)} & 0 \\ 0 & \sqrt{\lambda_2(U_D{}^2)} \end{bmatrix} \begin{bmatrix} \mathbf{n}_1 & \mathbf{n}_2 \end{bmatrix}^T = \begin{bmatrix} 4.6640 & 2.0608 \\ 2.0608 & 4.5555 \end{bmatrix}
$$

and

$$
R_O = Q U_D{}^{-1} = \begin{bmatrix} 0.9762 & 0.2169 \\ -0.2169 & 0.9762 \end{bmatrix}.
$$

Homogeneous Transformations: Let deformation gradient (Jacobian) matrix be only time-varying ($Q = Q(t)$), then continuum deformation is called *homogeneous transformation* or *homogeneous deformation*. By considering Eq. A.1, homogeneous transformation is expressed by

$$
r(t) = Q(t)R + D(t). \tag{A.14}
$$

where $D(t)$ is called *rigid body displacement vector*.

Appendix B
Notions of Graph Theory

A graph $G = G(V, E)$ consists of set of nodes

$$V = \{1,\ 2, \ldots,\ N\} \tag{B.1}$$

and set of edges $E \subset V \times V$. Two vertices $i, j \in V$ that share a common edge are called *adjacent*. An edge that is ended to a vertex is called *incident* to that vertex. An edge connecting a vertex j to a vertex i is denoted by $(j, i) \in E$. A graph is called a *null graph*, if $E = \emptyset$. The set

$$N_i = \{j | (j, i) \in V\} \tag{B.2}$$

is called *in-neighbor set of the node i*. A graph is called *simple*, if it does not have a loop or multiple edges. Total number of edges that are incident to a vertex i is called *degree* of the vertex i and denoted by d_i $(d_i = |N_i|)$.

Example. The graph shown in Fig. B.1 has the following properties:

- $V = \{1,\ 2, \ldots, 7\}$.
- $E = \{(3,4),\ (4,3),\ (1,4),\ (4,1),\ (1,5),\ (5,1),\ (6,7),\ (7,6),\ (4,7),\ (7,4),\ (5,4),\ (7,5)\}$.
- $N_4 = \{3,\ 1,\ 5,\ 7\}$.
- $d_7 = 2$.
- The node 2 is isolated.

Handshaking Lemma: If a graph is undirected, sum of all node degrees is even,

$$\sum_{i \in V} = 2m,\ m \in \mathbb{Z}. \tag{B.3}$$

© Springer International Publishing AG 2016
H. Rastgoftar, *Continuum Deformation of Multi-Agent Systems*,
DOI 10.1007/978-3-319-41594-9

Fig. B.1 A typical graph

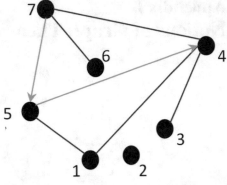

Fig. B.2 A typical graph

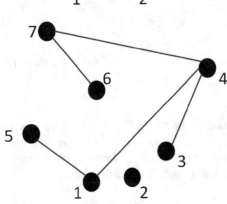

For example, in the graph shown in Fig. B.2,

$$d = \sum_{i=1}^{7} d_i = 2+0+1+1+1+1+2 = 8.$$

Remark. In a connected graph, number of nodes with odd degree is even (See Fig. B.2).

Subgraph: Let $G = (V, E)$ be a graph, then, $G_s = (V_s, E_s)$ is a subgraph of G, if

$$V_s \subset V, \ E_s \subset E.$$

Complement of a Graph: Let $G = (V, E)$ be a simple graph, then, $\bar{G} = (V, E')$ is the complement of G, if

$$E' \bigcap E = \emptyset$$
$$E' \bigcup E = V \times V \setminus \{(i, i) | i \in V\}.$$

Fig. B.3 Graph complement

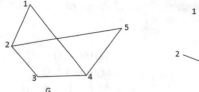 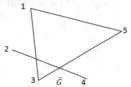

In Fig. B.3 a graph G and its complement are shown.

Complete Graph: A simple undirected graph is called *complete*, if every two different nodes is connected. For a complete graph,

$$|E| = \frac{N(N-1)}{2}.$$

Cycle C_n: A connected graph consisting of n node is called *cycle* C_n, if every node has the degree 2.

Path P_n: A path P_n consisting of n nodes is obtained by removing one edge from the circle C_n.

Walk: A walk in a graph $G = (V,E)$ is a sequence of edges in which every two consecutive edges share a common node. For example

$$(i_2, i_1),\ (i_3, i_2), \ldots,\ (i_{n-1}, i_{n-2}),\ (i_n, i_{n-1}) \tag{B.4}$$

represents a walk.

Trail: A walk is called a *trail*, if all edges are distinct.

Adjacency Matrix: For a graph $G = (V,E)$, the adjacency matrix $A = [A_{ij}] \in \mathbb{R}^{N \times N}$ is defined as follows:

$$A_{ij} = \begin{cases} Total\ number\ of\ edges & if\ (j,i) \in E \\ 0 & else. \end{cases}.$$

Remark. Adjacency matrix of a simple graph is binary with diagonal entries that are all 0.

Remark. Adjacency matrix of a simple undirected graph is symmetric.

Example. The adjacency matrix of the graph shown in Fig. B.1 becomes as follows:

$$A = \begin{bmatrix} 0 & 0 & 0 & 1 & 1 & 0 & 0 \\ 0 & 0 & 0 & 0 & 0 & 0 & 0 \\ 0 & 0 & 0 & 1 & 0 & 0 & 0 \\ 1 & 0 & 1 & 0 & 1 & 0 & 1 \\ 1 & 0 & 0 & 0 & 0 & 0 & 1 \\ 0 & 0 & 0 & 0 & 0 & 0 & 1 \\ 0 & 0 & 0 & 1 & 0 & 1 & 0 \end{bmatrix}.$$

Connectivity: Two nodes in a graph $G = (V, E)$ is called *connected*, if there exists a path from i to j. An undirected graph is *connected*, if every two nodes are connected. A digraph is called *weakly connected*, if replacing directed edges by undirected edges yields a connected graph. Two vertices i and j in the digraph G is called *connected*, if there exists a directed path from i to j, and one from j to i. A digraph G is called *strongly connected*, if every two nodes of G is connected.

Graph Laplacian: Given adjacency matrix A of a graph G, the graph Laplacian matrix is defined as follows:

$$L_{ij} = \begin{cases} 1 & \text{if } j = i \wedge d_i \neq 0 \\ \dfrac{-A_{ij}}{\sum_{j=1}^{N} A_{ij}} & \text{if } j \neq i \wedge d_i \neq 0, \\ 0 & \text{else} \end{cases} \tag{B.5}$$

Remark. If an undirected graph is connected, or a digraph is strongly connected, then the first eigenvalue of the Laplacian L is 0 and the remaining eigenvalues are all positive.

References

1. Ajorlou, A., Momeni, A, & Aghdam, A. G. (2009). Connectivity preservation in a network of single integrator agents. In *Proceedings of the 48th IEEE Conference on Decision and control, 2009 held jointly with the 2009 28th Chinese control conference. CDC/CCC 2009* (pp. 7061–7067). IEEE.
2. Ajorlou, A., Momeni, A., & Aghdam, A. G. (2010). A class of bounded distributed control strategies for connectivity preservation in multi-agent systems. *IEEE Transactions on Automatic Control, 55*(12), 2828–2833.
3. Alam, S. M. S., Natarajan, B., & Pahwa, A. (2015). Agent based optimally weighted Kalman consensus filter over a lossy network. In *2015 IEEE global communications conference (GLOBECOM)* (pp. 1–6). IEEE.
4. Antonelli, G., Arrichiello, F., & Chiaverini, S. (2006). Experiments of formation control with collisions avoidance using the null-space-based behavioral control. In *14th Mediterranean conference on control and automation, 2006. MED'06* (pp. 1–6). IEEE.
5. Antonelli, G., Arrichiello, F., & Chiaverini, S. (2009). Experiments of formation control with multirobot systems using the null-space-based behavioral control. *IEEE Transactions on Control Systems Technology, 17*(5), 1173–1182.
6. Bahceci, E., Soysal, O., & Sahin, E. (2003). A review: pattern formation and adaptation in multi-robot systems. *Robotics Institute, Carnegie Mellon University, Pittsburgh, PA.* Technical Report CMU-RI-TR-03-43.
7. Balch, T., & Arkin, R. C. (1998). Behavior-based formation control for multirobot teams. *IEEE Transactions on Robotics and Automation, 14*(6), 926–939.
8. Bazoula, A., Djouadi, M. S., & Maaref, H. (2008). Formation control of multi-robots via fuzzy logic technique. *International Journal of Computers, Communications Control, 3*(3).
9. Berman, A., & Plemmons, R. J. (1979). Nonnegative matrices. *The Mathematical Sciences, Classics in Applied Mathematics, 9.*
10. Binetti, G., Davoudi, A., Lewis, F. L., Naso, D., & Turchiano, B. (2014). Distributed consensus-based economic dispatch with transmission losses. *IEEE Transactions on Power Systems, 29*(4), 1711–1720.
11. Biswas, A., Thompson, D., He, W., Deng, Q., Chen, C.-M., Shenk, H.-W., et al. (2015). An uncertainty-driven approach to vortex analysis using oracle consensus and spatial proximity. In *2015 IEEE pacific visualization symposium (PacificVis)* (pp. 223–230). IEEE.
12. Bliman, P.-A., & Ferrari-Trecate, G. (2008). Average consensus problems in networks of agents with delayed communications. *Automatica, 44*(8), 1985–1995.

© Springer International Publishing AG 2016
H. Rastgoftar, *Continuum Deformation of Multi-Agent Systems*,
DOI 10.1007/978-3-319-41594-9

13. Bouteraa, Y., Ghommam, J., & Derbel, N. (2011). Coordinated backstepping control of multiple robot system of the leader-follower structure. In *2011 8th international multi-conference on systems, signals and devices (SSD)* (pp. 1–5). IEEE.

14. Cao, Y., & Ren, W. (2009). Containment control with multiple stationary or dynamic leaders under a directed interaction graph. In *Proceedings of the 48th IEEE conference on decision and control, 2009 held jointly with the 2009 28th Chinese control conference. CDC/CCC 2009* (pp. 3014–3019). IEEE.

15. Cao, Y., Stuart, D., Ren, W., & Meng, Z. (2011). Distributed containment control for multiple autonomous vehicles with double-integrator dynamics: Algorithms and experiments. *IEEE Transactions on Control Systems Technology, 19*(4), 929–938.

16. Carli, R., & Zampieri, S. (2014). Network clock synchronization based on the second-order linear consensus algorithm. *IEEE Transactions on Automatic Control, 59*(2), 409–422.

17. Chen, C.-T. (1995). *Linear system theory and design*. Oxford: Oxford University Press, Inc.

18. Chen, X., & Serrani, A. (2007). Smith predictors in nonlinear systems-application to ISS-based leader/follower trailing control. In *American control conference, 2007. ACC'07* (pp. 4506–4511). IEEE.

19. Chen, Y., & Wang, Z. (2005). Formation control: a review and a new consideration. In *2005 IEEE/RSJ international conference on intelligent robots and systems, 2005. (IROS 2005)* (pp. 3181–3186). IEEE.

20. Chen, C. L. P., Wen, G.-X., Liu, Y.-J., & Wang, F.-Y. (2014). Adaptive consensus control for a class of nonlinear multiagent time-delay systems using neural networks. *IEEE Transactions on Neural Networks and Learning Systems, 25*(6), 1217–1226.

21. Cheok, K. C., Smid, G.-E., Kobayashi, K., Overholt, J. L., & Lescoe, P. (1997). A fuzzy logic intelligent control system paradigm for an in-line-of-sight leader-following HMMWV. *Journal of Robotic Systems, 14*(6), 407–420.

22. Consolini, L., Morbidi, F., Prattichizzo, D., & Tosques, M. (2008). Leader–follower formation control of nonholonomic mobile robots with input constraints. *Automatica, 44*(5), 1343–1349.

23. Cortes, J., Martinez, S., Karatas, T., & Bullo, F. (2002). Coverage control for mobile sensing networks. In *IEEE international conference on robotics and automation, 2002. Proceedings. ICRA'02* (Vol. 2, pp. 1327–1332). IEEE.

24. Cui, R., Ge, S. S., Ee How, B. V., & Choo, Y. S. (2010). Leader–follower formation control of underactuated autonomous underwater vehicles. *Ocean Engineering, 37*(17), 1491–1502.

25. Desai, J. P., Ostrowski, J., & Kumar, V. (1998). Controlling formations of multiple mobile robots. In *1998 IEEE international conference on robotics and automation, 1998. Proceedings* (Vol. 4, pp. 2864–2869). IEEE.

26. Dierks, T., & Jagannathan, S (2007). Control of nonholonomic mobile robot formations: backstepping kinematics into dynamics. In *IEEE international conference on control applications, 2007. CCA 2007* (pp. 94–99).

27. Dierks, T., & Jagannathan, S. (2009). Neural network control of mobile robot formations using rise feedback. *IEEE Transactions on Systems, Man, and Cybernetics, Part B: Cybernetics, 39*(2), 332–347.

28. Dierks, T., & Jagannathan, S. (2010). Neural network output feedback control of robot formations. *IEEE Transactions on Systems, Man, and Cybernetics, Part B: Cybernetics, 40*(2), 383–399.

29. Dierks, T., Brenner, B., & Jagannathan, S. (2013). Neural network-based optimal control of mobile robot formations with reduced information exchange. *IEEE Transactions on Control Systems Technology, 21*(4), 1407–1415.

30. Dimarogonas, D. V., & Johansson, K. H. (2010). Stability analysis for multi-agent systems using the incidence matrix: quantized communication and formation control. *Automatica, 46*(4), 695–700.

31. Dong, X., Shi, Z., Ren, Z., & Zhong, Y. (2015). Output formation-containment control for high-order swarm systems with directed topologies. In *The 27th Chinese control and decision conference (2015 CCDC)* (pp. 36–43). IEEE.

32. Edwards, D. B., Bean, T. A., Odell, D. L., & Anderson, M. J. (2004). *A leader-follower algorithm for multiple AUV formations.* IEEE.
33. Fazelinia, H., Sipahi, R., & Olgac, N. (2007). Stability robustness analysis of multiple time-delayed systems using building block concept. *IEEE Transactions on Automatic Control, 52*(5), 799–810.
34. Ferrari-Trecate, G., Egerstedt, M., Buffa, A., & Ji, M. (2006). Laplacian sheep: A hybrid, stop-go policy for leader-based containment control. In *Hybrid systems: Computation and control* (pp. 212–226). Berlin: Springer.
35. Fridman, E. (2001). New Lyapunov–Krasovskii functionals for stability of linear retarded and neutral type systems. *Systems Control Letters, 43*(4), 309–319 (2001)
36. Frihauf, P., & Krstic, M. (2010). Multi-agent deployment to a family of planar arcs. In *American control conference (ACC), 2010*(pp. 4109–4114). IEEE.
37. Frihauf, P., & Krstic, M. (2011). Leader-enabled deployment onto planar curves: A PDE-based approach. *IEEE Transactions on Automatic Control, 56*(8), 1791–1806.
38. Gamage, G. W., Mann, G. K., & Gosine, R. G. (2010). Leader follower based formation control strategies for nonholonomic mobile robots: Design, implementation and experimental validation. In *American control conference (ACC), 2010* (pp. 224–229). IEEE..
39. Gayme, D. F., & Chakrabortty, A. (2012). Shaping power system inter-area oscillations through control loops of grid integrated wind farms. In *2012 IEEE 51st annual conference on decision and control (CDC)* (pp. 5004–5009). IEEE.
40. Gazi, V., & Passino, K. M. (2011). *Swarm stability and optimization.* New York: Springer Science Business Media.
41. Gerdes, J. C., & Rossetter, E. J. (2001). A unified approach to driver assistance systems based on artificial potential fields. *Journal of Dynamic Systems, Measurement, and Control, 123*(3), 431–438.
42. Ghods, N., & Krstic, M. (2012). Multiagent deployment over a source. *IEEE Transactions on Control Systems Technology, 20*(1), 277–285.
43. Ghommam, J., Saad, M., & Mnif, F. (2010). Robust adaptive formation control of fully actuated marine vessels using local potential functions. In *2010 IEEE international conference on robotics and automation (ICRA)* (pp. 3001–3007). IEEE.
44. Ghommam, J., Mehrjerdi, H., & Saad, M. (2011). Leader-follower formation control of nonholonomic robots with fuzzy logic based approach for obstacle avoidance. In *2011 IEEE/RSJ international conference on intelligent robots and systems (IROS)* (pp. 2340–2345). IEEE.
45. Haghpanahi, M., Sameni, R., & Borkholder, D. A. (2014). Scoring consensus of multiple ecg annotators by optimal sequence alignment. In *2014 36th annual international conference of the IEEE engineering in medicine and biology society* (pp. 1855–1859). IEEE.
46. Han, T., Lin, Z., & Fu, M. (2015). Three-dimensional formation merging control under directed and switching topologies. *Automatica, 58*, 99–105.
47. Hao, H., & Barooah, P. (2010). Control of large 1−d networks of double integrator agents: role of heterogeneity and asymmetry on stability margin. In *49th IEEE conference on decision and control (CDC)* (pp. 7395–7400). IEEE.
48. He, X., Wang, Q., & Yu, W. (2014). Finite-time containment control for second-order multiagent systems under directed topology. *IEEE Transactions on Circuits and Systems II: Express Briefs, 61*(8), 619–623.
49. Hettiarachchi, S., Spears, W. M., Varghese, B., & McKee, G. (2009). A review and implementation of swarm pattern formation and transformation models. *International Journal of Intelligent Computing and Cybernetics, 2*(4), 786–817.
50. Hu, J., & Cao, J. (2015). Hierarchical cooperative control for multiagent systems with switching directed topologies. *IEEE Transactions on Neural Networks and Learning Systems, 26*(10), 2453–2463.
51. Hu, Y., Lam, J., & Liang, J. (2013). Consensus control of multi-agent systems with missing data in actuators and Markovian communication failure. *International Journal of Systems Science, 44*(10), 1867–1878.

52. Huang, T., & Chen, X. (2008). A geometric method of swarm robot formation controls. In *7th world congress on intelligent control and automation, 2008. WCICA 2008* (pp. 3202–3206). IEEE.
53. Hui, Q., Haddad, W. M., & Bhat, S. P. (2010). On robust control algorithms for nonlinear network consensus protocols. *International Journal of Robust and Nonlinear Control, 20*(3), 269–284.
54. Ji, M., Ferrari-Trecate, G., Egerstedt, M., & Buffa, A. (2008). Containment control in mobile networks. *IEEE Transactions on Automatic Control, 53*(8), 1972–1975.
55. Jia, Q., & Li, G. (2007). Formation control and obstacle avoidance algorithm of multiple autonomous underwater vehicles (AUVS) based on potential function and behavior rules. In *2007 IEEE international conference on automation and logistics* (pp. 569–573). IEEE.
56. Jia, B., Pham, K. D., Blasch, E., Shen, D., Wang, Z., & Chen, G. (2014). Cooperative space object tracking using consensus-based filters. In *2014 17th international conference on information fusion (FUSION)* (pp. 1–8). IEEE.
57. Kar, S., & Moura, J. M. F. (2009). Distributed consensus algorithms in sensor networks with imperfect communication: Link failures and channel noise. *IEEE Transactions on Signal Processing, 57*(1), 355–369.
58. Kharitonov, V. L., & Zhabko, A. P. (2003). Lyapunov–Krasovskii approach to the robust stability analysis of time-delay systems. *Automatica, 39*(1), 15–20.
59. Khoo, S., Xie, L., Yu, Z., & Man, Z. (2008). Finite-time consensus algorithm of multi-agent networks. In *10th international conference on control, automation, robotics and vision, 2008. ICARCV 2008* (pp. 916–920). IEEE.
60. Khoo, S., Yin, J., Wang, B., Zhao, S., & Man, Z. (2011). Adaptive data based neural network leader-follower control of multi-agent networks. In *IECON 2011-37th annual conference on IEEE industrial electronics society* (pp. 3924–3929). IEEE.
61. Kim, J., Kim, K.-D., Natarajan, V., Kelly, S. D., & Bentsman, J. (2008). PDE-based model reference adaptive control of uncertain heterogeneous multiagent networks. *Nonlinear Analysis: Hybrid Systems, 2*(4), 1152–1167.
62. Kloetzer, M., & Belta, C. (2007). Temporal logic planning and control of robotic swarms by hierarchical abstractions. *IEEE Transactions on Robotics, 23*(2), 320–330.
63. Kvinto, Y. I., & Parsegov, S. E. (2012). Equidistant arrangement of agents on line: Analysis of the algorithm and its generalization. *Automation and Remote Control, 73*(11), 1784–1793.
64. Lai, W. M., Rubin, D. H., Rubin, D., & Krempl, E. (2009). *Introduction to continuum mechanics*. Oxford: Butterworth-Heinemann.
65. Lestas, I., & Vinnicombe, G. (2010). Heterogeneity and scalability in group agreement protocols: Beyond small gain and passivity approaches. *Automatica, 46*(7), 1141–1151.
66. Li, Q., & Jiang, Z.-P. (2008). Formation tracking control of unicycle teams with collision avoidance. In *47th IEEE conference on decision and control, 2008. CDC 2008* (pp. 496–501). IEEE.
67. Li, X., Xiao, J., & Cai, Z. (2005). Backstepping based multiple mobile robots formation control. In *2005 IEEE/RSJ international conference on intelligent robots and systems, 2005.(IROS 2005)* (pp. 887–892). IEEE.
68. Li, H., Djouadi, S., & Tomsovic, K. (2012). Flocking generators: A PdE framework for stability of smart grids with communications. In *2012 IEEE third international conference on smart grid communications (SmartGridComm)* (pp. 540–545). IEEE.
69. Li, H., Liao, X., Lei, X., Huang, T., & Zhu, W. (2013). Second-order consensus seeking in multi-agent systems with nonlinear dynamics over random switching directed networks. *IEEE Transactions on Circuits and Systems I: Regular Papers, 60*(6), 1595–1607.
70. Li, S., Feng, G., Luo, X., & Guan, X. (2015). Output consensus of heterogeneous linear discrete-time multiagent systems with structural uncertainties. *IEEE Transactions on Cybernetics, 45*(12), 2868–2879.
71. Lin, P., & Jia, Y. (2010). Consensus of a class of second-order multi-agent systems with time-delay and jointly-connected topologies. *IEEE Transactions on Automatic Control, 55*(3), 778–784.

72. Lin, P., Jia, Y., & Li, L. (2008). Distributed robust consensus control in directed networks of agents with time-delay. *Systems Control Letters, 57*(8), 643–653.
73. Lin, Z., Ding, W., Yan, G., Yu, C., & Giua, A. (2013). Leader–follower formation via complex Laplacian. *Automatica, 49*(6), 1900–1906.
74. Liu, B., & Chen, T. (2008). Consensus in networks of multiagents with cooperation and competition via stochastically switching topologies. *IEEE Transactions on Neural Networks, 19*(11), 1967–1973.
75. Liu, C.-L., & Liu, F. (2014). Asynchronously compensated consensus algorithm for discrete-time second-order multi-agent systems under communication delay. *IET Control Theory Applications, 8*(17), 2004–2012.
76. Liu, B., Zhang, R., & Shi, C. (2006). Formation control of multiple behavior-based robots. In *2006 international conference on computational intelligence and security* (Vol. 1, pp. 544–547). IEEE.
77. Liu, W., Gu, W., Sheng, W., Meng, X., Wu, Z., & Chen, W. (2014). Decentralized multi-agent system-based cooperative frequency control for autonomous microgrids with communication constraints. *IEEE Transactions on Sustainable Energy, 5*(2), 446–456 (2014)
78. Lopez-Martinez, M., Delvenne, J.-C., & Blondel, V. D. (2012). Optimal sampling time for consensus in time-delayed networked systems. *IET Control Theory Applications, 6*(15), 2467–2476.
79. Low, C. B., & Ng, Q. S. (2011). A flexible virtual structure formation keeping control for fixed-wing UAVS. In *2011 9th IEEE international conference on control and automation (ICCA)* (pp. 621–626). IEEE.
80. Magar, K. T., Balas, M. J., & Gayme, D. F. (2014). Adaptive control of inter-area oscillations in wind-integrated power systems using distributed parameter control methods. In *American control conference (ACC), 2014* (pp. 903–907). IEEE.
81. Mariottini, G. L., Morbidi, F., Prattichizzo, D., Valk, N. V., Michael, N., Pappas, G., et al. (2009). Vision-based localization for leader–follower formation control. *IEEE Transactions on Robotics, 25*(6), 1431–1438.
82. Mehrjerdi, H., Ghommam, J., & Saad, M. (2011). Nonlinear coordination control for a group of mobile robots using a virtual structure. *Mechatronics, 21*(7), 1147–1155.
83. Meurer, T., & Krstic, M. (2010). Nonlinear PDE-based motion planning for the formation control of mobile agents. In *Proceedings of the (CD–ROM) 8th IFAC symposium nonlinear control systems (NOLCOS 2010), Bologna (I)* (pp. 599–604).
84. Miyasato, Y. (2010). Adaptive H formation control for Euler-Lagrange systems. In *2010 49th IEEE conference on decision and control (CDC)* (pp. 2614–2619). IEEE.
85. Miyasato, Y. (2011). Adaptive H formation control for Euler-Lagrange systems by utilizing neural network approximators. In *American control conference (ACC), 2011* (pp. 1753–1758). IEEE.
86. Motee, N., Jadbabaie, A., & Pappas, G. (2010). Path planning for multiple robots: An alternative duality approach. In *American control conference (ACC), 2010* (pp. 1611–1616). IEEE.
87. Moura, S., Bendtsen, J., & Ruiz, V. (2013). Observer design for boundary coupled PDES: Application to thermostatically controlled loads in smart grids. In *52nd IEEE conference on decision and control* (pp. 6286–6291). IEEE.
88. Mu, X., & Zheng, B. (2015). Containment control of second-order discrete-time multi-agent systems with Markovian missing data. *IET Control Theory Applications, 9*(8), 1229–1237.
89. Münz, U., Papachristodoulou, A., & Allgöwer, F. (2010). Delay robustness in consensus problems. *Automatica, 46*(8), 1252–1265.
90. Munz, U., Papachristodoulou, A., & Allgower, F. (2011). Robust consensus controller design for nonlinear relative degree two multi-agent systems with communication constraints. *IEEE Transactions on Automatic Control, 56*(1), 145–151.
91. Murray, R. M. (2007). Recent research in cooperative control of multivehicle systems. *Journal of Dynamic Systems, Measurement, and Control, 129*(5), 571–583.

92. Nathan, D. M., Buse, J. B., Davidson, M. B., Ferrannini, E., Holman, R. R., Sherwin, R., et al. (2009). Medical management of hyperglycemia in type 2 diabetes: a consensus algorithm for the initiation and adjustment of therapy a consensus statement of the American diabetes association and the European association for the study of diabetes. *Diabetes Care, 32*(1), 193–203.

93. Olfati-Saber, R., & Murray, R. M. (2004). Consensus problems in networks of agents with switching topology and time-delays. *IEEE Transactions on Automatic Control, 49*(9), 1520–1533.

94. Olgac, N., & Sipahi, R. (2016). A practical method for analyzing the stability of neutral type LTI-time delayed systems. *Automatica, 40*(5), 847–853.

95. Omidi, E., & Mahmoodi, S. N. (2016). Robust optimal consensus state estimator for a piezoactive distributed parameter system. *Journal of Dynamic Systems, Measurement, and Control, 138*(9), 091011.

96. Panigrahi, N., & Khilar, P. M. (2015). Optimal consensus-based clock synchronisation algorithm in wireless sensor network by selective averaging. *IET Wireless Sensor Systems, 5*(3), 166–174.

97. Parsegov, S., Polyakov, A., & Shcherbakov, P. (2012). Nonlinear fixed-time control protocol for uniform allocation of agents on a segment. In *2012 IEEE 51st IEEE conference on decision and control (CDC)* (pp. 7732–7737). IEEE.

98. Patterson, S., Bamieh, B., & Abbadi, A. E. (2007). Distributed average consensus with stochastic communication failures. In *2007 46th IEEE conference on decision and control* (pp. 4215–4220). IEEE.

99. Pepe, P., & Jiang, Z.-P. (2006). A Lyapunov–Krasovskii methodology for ISS and IISS of time-delay systems. *Systems Control Letters, 55*(12), 1006–1014.

100. Pereira, A. R., Hsu, L., & Ortega, R. (2009). Globally stable adaptive formation control of Euler-Lagrange agents via potential functions. In *American control conference, 2009. ACC'09* (pp. 2606–2611). IEEE.

101. Qin, J., Ma, Q., Zheng, W. X., & Gao, H. (2015). H group consensus for clusters of agents with model uncertainty and external disturbance. In *2015 54th IEEE conference on decision and control (CDC)* (pp. 2841–2846). IEEE.

102. Qu, Z. (2009). *Cooperative control of dynamical systems: Applications to autonomous vehicles*. New York: Springer Science Business Media.

103. Ranjbar-Sahraei, B., Shabaninia, F., Nemati, A., & Stan, S.-D. (2012). A novel robust decentralized adaptive fuzzy control for swarm formation of multiagent systems. *IEEE Transactions on Industrial Electronics, 59*(8), 3124–3134.

104. Rastgoftar, H. (2013). *Planning and control of swarm motion as continua*. Ph.D. Thesis, University of Central Florida Orlando, Florida.

105. Rastgoftar, H., & Atkins, E. M. (2017). Continuum deformation of multi-agent systems under directed communication topologies. *Journal of Dynamic Systems, Measurement, and Control, 139*(1), 011002.

106. Rastgoftar, H., & Jayasuriya, S. (2013). Distributed control of swarm motions as continua using homogeneous maps and agent triangulation. In *2013 European control conference (ECC)* (pp. 2824–2830). IEEE.

107. Rastgoftar, H., & Jayasuriya, S. (2013). Multi-agent deployment based on homogeneous maps and a special inter-agent communication protocol. *Mechatronic Systems, 1*, 569–576.

108. Rastgoftar, H., & Jayasuriya, S. (2013). Preserving stability under communication delays in multi agent systems. In *ASME 2013 dynamic systems and control conference* (pp. V002T21A002–V002T21A002). American Society of Mechanical Engineers.

109. Rastgoftar, H., & Jayasuriya, S. (2014). Alignment as biological inspiration for control of multi agent systems. In *ASME 2014 dynamic systems and control conference* (pp. V001T05A003–V001T05A003). American Society of Mechanical Engineers.

110. Rastgoftar, H., & Jayasuriya, S. (2014). A continuum based approach for multi agent systems under local inter-agent communication. In *American control conference (ACC), 2014* (pp. 825–830). IEEE.

111. Rastgoftar, H., & Jayasuriya, S. (2014). Evolution of multi-agent systems as continua. *Journal of Dynamic Systems, Measurement, and Control, 136*(4), 041014.

112. Rastgoftar, H., & Jayasuriya, S. (2015). An alignment strategy for evolution of multi-agent systems. *Journal of Dynamic Systems, Measurement, and Control, 137*(2), 021009. American Society of Mechanical Engineers.

113. Rastgoftar, H., & Jayasuriya, S. (2015). Swarm motion as particles of a continuum with communication delays. *Journal of Dynamic Systems, Measurement, and Control, 137*(11), 111008.

114. Rastgoftar, H., & Atkins, E. M. (2017). Continuum deformation of multi-agent systems under directed communication topologies. *Journal of Dynamic Systems, Measurement, and Control, 139*(1), 011002.

115. Rastgoftar, H., Kwatny, H. G., & Atkins, E. M. (2017). Asymptotic tracking and robustness of MAS transitions under a new communication topology. *IEEE Transactions on Automation Science and Engineering*, pp(99).

116. Ravazzi, C., Fosson, S. M., & Magli, E. (2015). Distributed iterative thresholding for-regularized linear inverse problems. *IEEE Transactions on Information Theory, 61*(4), 2081–2100.

117. Ren, W. (2007). Consensus strategies for cooperative control of vehicle formations. *Control Theory Applications, IET, 1*(2), 505–512.

118. Ren, W., & Beard, R. W. (2004). Formation feedback control for multiple spacecraft via virtual structures. In *IEE proceedings control theory and applications* (Vol. 151, pp. 357–368). IET.

119. Ren, W., & Beard, R. W. (2008). Consensus algorithms for double-integrator dynamics. *Distributed consensus in multi-vehicle cooperative control: theory and applications* (pp. 77–104). London: Springer.

120. Ren, J., McIsaac, K., et al. (2003). A hybrid-systems approach to potential field navigation for a multi-robot team. In *IEEE international conference on robotics and automation, 2003. Proceedings. ICRA'03* (Vol. 3, pp. 3875–3880). IEEE.

121. Rezaee, H., & Abdollahi, F. (2011). Mobile robots cooperative control and obstacle avoidance using potential field. In *2011 IEEE/ASME international conference on advanced intelligent mechatronics (AIM)* (pp. 61–66). IEEE.

122. Rong, L., Lu, J., Xu, S., & Chu, Y. (2014). Reference model-based containment control of multi-agent systems with higher-order dynamics. *IET Control Theory Applications, 8*(10), 796–802.

123. Rossetter, E. J., Switkes, J. P., & Gerdes, J. C. (2004). Experimental validation of the potential field lanekeeping system. *International journal of automotive technology, 5*(2), 95–108.

124. Roussos, G., & Kyriakopoulos, K. J. (2010). Completely decentralised navigation of multiple unicycle agents with prioritisation and fault tolerance. In *2010 49th IEEE conference on decision and control (CDC)* (pp. 1372–1377). IEEE.

125. Sabattini, L., Secchi, C., Fantuzzi, C., de Macedo Possamai, D., et al. (2010). Tracking of closed-curve trajectories for multi-robot systems. In *IROS* (pp. 6089–6094).

126. Sahraei, B. R., & Shabaninia, F. (2010). A robust h control design for swarm formation control of multi-agent systems: a decentralized adaptive fuzzy approach. In *2010 3rd international symposium on resilient control systems (ISRCS)* (pp. 79–84). IEEE.

127. Sepulchre, R., Paley, D., Leonard, N. E., et al. (2008). Stabilization of planar collective motion with limited communication. *IEEE Transactions on Automatic Control, 53*(3), 706–719.

128. Shao, J., Xie, G., & Wang, L. (2007). Leader-following formation control of multiple mobile vehicles. *IET Control Theory Applications, 1*(2), 545–552.

129. Shi-Cai, L., Da-Long, T, & Guang-Jun, L. (2007). Robust leader-follower formation control of mobile robots based on a second order kinematics model. *Acta Automatica Sinica, 33*(9), 947–955.

130. Simon, D. (2006). *Optimal state estimation: Kalman, H infinity, and nonlinear approaches.* New York: Wiley.

131. Sipahi, R., & Olgac, N. (2006). Complete stability analysis of neutral-type first order two-time-delay systems with cross-talking delays. *SIAM journal on control and optimization, 45*(3), 957–971.

132. Sisto, M., & Gu, D. (2006). A fuzzy leader-follower approach to formation control of multiple mobile robots. In *2006 IEEE/RSJ international conference on intelligent robots and systems* (pp. 2515–2520). IEEE.

133. Song, H., Yu, L., & Zhang, W.-A. (2014). Distributed consensus-based Kalman filtering in sensor networks with quantised communications and random sensor failures. *IET Signal Processing, 8*(2), 107–118.

134. Srinivasan, S., & Ayyagari, R. (2010). Consensus algorithm for robotic agents over packet dropping links. In *2010 3rd international conference on biomedical engineering and informatics (BMEI)* (Vol. 6, pp. 2636–2640). IEEE.

135. Su, H., & Chen, M. Z. Q. (2015). Multi-agent containment control with input saturation on switching topologies. *IET Control Theory Applications, 9*(3), 399–409.

136. Tahbaz-Salehi, A., & Jadbabaie, A. (2010). Consensus over ergodic stationary graph processes. *IEEE Transactions on Automatic Control, 55*(1), 225–230.

137. Tan, C., & Liu, G.-P. (2013). Consensus of discrete-time linear networked multi-agent systems with communication delays. *IEEE Transactions on Automatic Control, 58*(11), 2962–2968.

138. Urcola, P., Riazuelo, L., Lazaro, M. T., & Montano, L. (2008). Cooperative navigation using environment compliant robot formations. In *IEEE/RSJ international conference on intelligent robots and systems, 2008. IROS 2008*, (pp. 2789–2794). IEEE.

139. Vanka, S., Gupta, V., & Haenggi, M. (2009). On consensus over stochastically switching directed topologies. In *American control conference, 2009. ACC'09.* (pp. 4531–4536). IEEE.

140. Vassilaras, S., Vogiatzis, D., & Yovanof, G. S. (2066). Security and cooperation in clustered mobile ad hoc networks with centralized supervision. *IEEE Journal on Selected Areas in Communications, 24*(2), 329–342

141. Vidal, R., Shakernia, O., & Sastry, S. (2004). Following the flock [formation control]. *IEEE Robotics Automation Magazine, 11*(4), 14–20.

142. Wagner, I. A., & Bruckstein, A. M. (1997). Row straightening via local interactions. *Circuits, Systems and Signal Processing, 16*(3), 287–305.

143. Wang, Z., & Gu, D. (2007). Distributed cohesion control for leader-follower flocking. In *IEEE international fuzzy systems conference, 2007. FUZZ-IEEE 2007* (pp. 1–6). IEEE.

144. Wang, S., & Schaub, H. (2011). Nonlinear feedback control of a spinning two-spacecraft coulomb virtual structure. *IEEE Transactions on Aerospace and Electronic Systems, 47*(3), 2055–2067 (2011)

145. Wang, L., Han, Z., & Lin, Z. (2012). Formation control of directed multi-agent networks based on complex Laplacian. In *2012 IEEE 51st annual conference on decision and control (CDC)* (pp. 5292–5297). IEEE.

146. Wang, X., Li, S., & Shi, P. (2014). Distributed finite-time containment control for double-integrator multiagent systems. *IEEE transactions on cybernetics, 44*(9), 1518–1528 (2014)

147. Wei, J., & Fang, H. (2014). Multi-agent consensus with time-varying delays and switching topologies. *Journal of Systems Engineering and Electronics, 25*(3), 489–495.

148. Weijun, S., Rui, M., & Chongchong, Y. (2010). A study on soccer robot path planning with fuzzy artificial potential field. In *2010 international conference on computing, control and industrial engineering (CCIE)* (Vol. 1, pp. 386–390). IEEE.

149. Wen, G., Duan, Z., Li, Z., & Chen, G. (2012). Stochastic consensus in directed networks of agents with non-linear dynamics and repairable actuator failures. *IET Control Theory Applications, 6*(11), 1583–1593.

150. Wen, G., Duan, Z., Chen, G., & Yu, W. (2014). Consensus tracking of multi-agent systems with Lipschitz-type node dynamics and switching topologies. *IEEE Transactions on Circuits and Systems I: Regular Papers, 61*(2), 499–511.

151. Wen, G., Hu, G., Yu, W., & Chen, G. (2014). Distributed consensus of higher order multiagent systems with switching topologies. *IEEE Transactions on Circuits and Systems II: Express Briefs, 61*(5), 359–363.

152. Wen, G., Hu, G., Hu, J., Shi, X., & Chen, G. (2016). Frequency regulation of source-grid-load systems: A compound control strategy. *IEEE Transactions on Industrial Informatics, 12*(1), 69–78.

153. Wen, G., Zhao, Y., Duan, Z., Yu, W., & Chen, G. (2016). Containment of higher-order multi-leader multi-agent systems: a dynamic output approach. *IEEE Transactions on Automatic Control, 61*(4), 1135–1140.

154. Wolf, M. T., & Burdick, J. W. (2008). Artificial potential functions for highway driving with collision avoidance. In *IEEE international conference on robotics and automation, 2008. ICRA 2008* (pp. 3731–3736). IEEE.

155. Xi, J., Xu, Z., Liu, G., & Zhong, Y. (2013). Stable-protocol output consensus for high-order linear swarm systems with time-varying delays. *IET Control Theory Applications, 7*(7), 975–984.

156. Xi, J., Yu, Y., Liu, G., & Zhong, Y. (2014). Guaranteed-cost consensus for singular multi-agent systems with switching topologies. *IEEE Transactions on Circuits and Systems I: Regular Papers, 61*(5), 1531–1542.

157. Xin, M., Balakrishnan, S. N., & Pernicka, H. J. (2007). Multiple spacecraft formation control with OD method. *IET Control Theory Applications, 1*(2), 485–493.

158. Xu, Y., & Liu, W. (2011). Novel multiagent based load restoration algorithm for microgrids. *IEEE Transactions on Smart Grid, 2*(1), 152–161.

159. Xu, X., Xie, J., & Xie, K. (2006). Path planning and obstacle-avoidance for soccer robot based on artificial potential field and genetic algorithm. In *The sixth world congress on intelligent control and automation, 2006. WCICA 2006* (Vol. 1, pp. 3494–3498). IEEE.

160. Xu, Y., Liu, W., & Gong, J. (2011). Stable multi-agent-based load shedding algorithm for power systems. *IEEE Transactions on Power Systems, 26*(4), 2006–2014.

161. Xue, D., Yao, J., Wang, J., Guo, Y., & Han, X. (2013). Formation control of multi-agent systems with stochastic switching topology and time-varying communication delays. *IET Control Theory Applications, 7*(13), 1689–1698.

162. Yamashita, A., Fukuchi, M., Ota, J., Arai, T., & Asama, H. (2000). Motion planning for cooperative transportation of a large object by multiple mobile robots in a 3d environment. In *IEEE international conference on robotics and automation, 2000. Proceedings. ICRA'00* (Vol. 4, pp. 3144–3151). IEEE.

163. Yang, E., & Gu, D. (2007). Nonlinear formation-keeping and mooring control of multiple autonomous underwater vehicles. *IEEE/ASME Transactions on Mechatronics, 12*(2), 164–178.

164. Yang, E., Gu, D., & Hu, H. (2005). Improving the formation-keeping performance of multiple autonomous underwater robotic vehicles. In *2005 IEEE international conference mechatronics and automation* (Vol. 4, pp. 1890–1895). IEEE.

165. Yang, X.-X., Tang, G.-Y., Li, Y., & Wang, P.-D. (2012). Formation control for multiple autonomous agents based on virtual leader structure. In *2012 24th Chinese control and decision conference (CCDC)* (pp. 2833–2837). IEEE.

166. Yoo, S. J., Park, J. B., & Choi, Y. H. (2010). Adaptive formation tracking control of electrically driven multiple mobile robots. *IET Control Theory Applications, 4*(8), 1489–1500.

167. Yoshioka, C., & Namerikawa, T. (2008). Formation control of nonholonomic multi-vehicle systems based on virtual structure. In *17th IFAC world congress* (pp. 5149–5154).

168. Yu, W., & Chen, G. (2010). Robust adaptive flocking control of nonlinear multi-agent systems. In *2010 IEEE international symposium on computer-aided control system design (CACSD)* (pp. 363–367). IEEE.

169. Yu, H., Keshavamurthy, S., Sheorey, H. B. S., Nguyen, H., & Taylor, C. N. (2014). Uncertainty estimation for random sample consensus. In *2014 13th international conference on control automation robotics vision (ICARCV)* (pp. 395–400). IEEE.

170. Zhang, W., Guo, Y., Liu, H., Chen, Y. J., Wang, Z., & Mitola, III, J. (2015). Distributed consensus-based weight design for cooperative spectrum sensing. *IEEE Transactions on Parallel and Distributed Systems, 26*(1), 54–64.

171. Zhang, D., Jiang, J., Wang, L. Y., & Zhang, W. (2016). Robust and scalable management of power networks in dual-source trolleybus systems: A consensus control framework. *IEEE Transactions on Intelligent Transportation Systems, 17*(4), 1029–1038.

172. Zhao, J., Su, X., & Yan, J. (2009). A novel strategy for distributed multi-robot coordination in area exploration. In *International conference on measuring technology and mechatronics automation, 2009. ICMTMA'09* (Vol. 2, pp. 24–27). IEEE.

173. Zhao, D., Zou, T., Li, S., & Zhu, Q. (2012). Adaptive backstepping sliding mode control for leader–follower multi-agent systems. *IET Control Theory Applications, 6*(8), 1109–1117.

174. Zhao, W., Liu, M., Zhu, J., & Li, L. (2016). Fully decentralised multi-area dynamic economic dispatch for large-scale power systems via cutting plane consensus. *IET Generation, Transmission Distribution.*

175. Zheng, Y., & Wang, L. (2014). Containment control of heterogeneous multi-agent systems. *International Journal of Control, 87*(1), 1–8.

176. Zhou, F., & Wang, Z. (2015). Containment control of linear multi-agent systems with directed graphs and multiple leaders of time-varying bounded inputs. *IET Control Theory Applications, 9*(16), 2466–2473.

177. Zhou, S., & Li, T. (2005). Robust stabilization for delayed discrete-time fuzzy systems via basis-dependent Lyapunov–Krasovskii function. *Fuzzy Sets and Systems, 151*(1), 139–153.

178. Zuo, Z., Zhang, J., & Wang, Y. (2014). Distributed consensus of linear multi-agent systems with fault tolerant control protocols. In *2014 33rd Chinese control conference (CCC)* (pp. 1656–1661). IEEE.

Index

Printed in the United States
By Bookmasters